Threat Modeling Gameplay with EoP

A reference manual for spotting threats in software architecture

Brett Crawley

Threat Modeling Gameplay with EoP

Group Product Manager: Dhruv J. Kataria
Publishing Product Manager: Prachi Sawant
Book Project Manager: Srinidhi Ram
Senior Editor: Adrija Mitra
Technical Editor: Arjun Varma
Copy Editor: Safis Editing
Proofreader: Adrija Mitra
Indexer: Tejal Soni
Production Designer: Jyoti Kadam
Senior DevRel Marketing Executive: Marylou De Mello
DevRel Marketing Coordinator: Shruthi Shetty

First published: August 2024

Production reference: 1110724

Published by Packt Publishing Ltd.
Grosvenor House
11 St Paul's Square
Birmingham
B3 1RB, UK

ISBN 978-1-80461-897-4

www.packtpub.com

To all those trying to make us all a little safer by writing secure software, thanks and I hope this work helps to some extent.

– Brett Crawley

Foreword

Let me start by inviting you to join me in a little game. Please complete the sentence: "It was Colonel Mustard…" Anyone who's lived in an English-speaking country and has played Clue will respond with a variant of "in the library with the lead pipe." Answers spring to mind even 10 or 20 years after last playing the game.

Games teach. They teach because while we're having fun, we're open to learning. We're engaged with what's in front of us, and learning about it helps us both have fun and win!

That was the inspiration that led me to create Elevation of Privilege in 2010. I had created software to help people threat model, and it was tedious to use—the exact opposite of the fun that security experts have while threat modeling. I wanted to bring that fun to those less experienced in security. So, long story short, I created the game, and hundreds of thousands of copies have been produced and used by people all around the world to learn and encourage threat modeling.

Play-testing showed how powerful it could be. People were unable to be "wallflowers," sometimes to their temporary regret as it was their turn and they didn't know what to do. But the game framing gives permission to be playful, and with the hints on the cards, that is a powerful combination. And there's just something cool about going into a business meeting with a game. It's going to be different, and if you suspend your skepticism, what happens can be magical.

Or not? Sometimes people get caught up in not knowing what a card means. And that's why I'm so excited about the book you're holding. Brett has stepped up to create a manual and open the world of security to a whole new audience.

The design of the game involved some lucky choices, and luck favors the prepared mind. So, while a lifetime of playing games helped, there's an entire field of study of "games with a purpose" or "serious games." And Elevation of Privilege helped show that games can help us learn about or even deliver security. I encourage you to use the game, and this book, to empower those around you to deliver more secure systems.

Adam Shostack

Creator of EoP

Seattle, WA

March 2024

Contributors

About the author

Brett Crawley is a principal application security engineer, (ISC2) CISSP, CSSLP, and CCSP certified, the project lead on the OWASP Application Security Awareness Campaigns project, and the author of the OSTERING.com blog on security. He has published a Miro template for threat modeling with the Elevation of Privilege card game and also published the CAPEC S.T.R.I.D.E. mapping mind maps and other resources.

With over 10 years of application security experience and over 25 years of software engineering experience, he works with teams to define their security best practices and introduce security by design into their existing SDLC, and as part of this initiative, he trains teams in threat modeling because good design is of key importance.

He is also an advocate for using a data-driven approach to AppSec, to help identify the business-critical components, thereby optimizing the reduction of risk to the organization.

I'd like to thank my wife, Sabrina, for her help with the illustrations in the book, as well as putting up with me; my friend, Victoria, for her feedback on the first draft of the book; and my employer, Mimecast, for enabling our team to build a threat modeling program and train our engineering teams to threat model using Elevation of Privilege, which made me realize that a book like this one would be a helpful resource for people new to threat modeling.

I'd obviously like to thank Adam Shostack for inventing the Elevation of Privilege game (© 2010 Microsoft Corporation, licensed under the Creative Commons Attribution 3.0 United States license), which this book is written to support; Mark Vinkovits for his Privacy Extension (© 2018 LogMeIn, Inc. licensed under the Creative Commons Attribution 3.0 United States License); and Marko Hämäläinen, Laura Noukka, Hiski Ruhanen, Ilona Varis, and Antti Vähä-Sipilä of F-Secure Corporation for their Elevation of Privacy (© 2018 F-Secure Corporation licensed under the Creative Commons Attribution 4.0 International license), which covers T.R.I.M. (Transport, Retention/Removal, Inference, and Minimization). Without their efforts, I wouldn't have anything to write about.

I'd also like to thank the MITRE Corporation for their CAPEC and CWE resources, which I refer to throughout the book (© 2006–2024 The MITRE Corporation), and OWASP for the Application Security Verification Standard ASVS whose section numbers I reference.

About the reviewers

Zoe Braiterman is an entrepreneur, IT security consultant/researcher, and open source contributor. She is also an enthusiastic volunteer with organizations such as OWASP and AFCEA.

Michael Bernhardt is a seasoned security strategist and believes that a solid security culture is the essential glue for technological innovation and strong security. Throughout his more than 15 years in the profession, he has advised dozens of Fortune-500 SAP ERP customers and is currently helping Germany's second-largest telecommunication provider in their secure cloud transformation as head of product security. He is leading the Corporate Security Program Evolution Model (CSPEM) initiative, which brings along tools and concepts for the organizational transformation of security programs. Additionally, he is a founder of the OWASP Security Champions Manifesto and Threat Modeling Connect, and regularly shares his perspective at conferences and on blogs.

Security depends on the diverse background of its workforce to assure broad acceptance. Failure in doing so would be fatal considering the magnitude of digital transformation.

Starting Threat Modeling more than 15 years ago, I very quickly understood its power to establish a collaborative exchange with diverse teams and raise security awareness. However, getting the first workshop done is one of the biggest challenges for new joiners. This is where "Threat Modeling Gameplay with EoP" steps in and brings structured real-world examples for guidance. I'm confident it will also aid you as you begin your own threat modeling journey!

Table of Contents

3

Tampering   37

4

Repudiation   63

7

Elevation of Privilege   137

8

Privacy   157

9

Transfer   173

10

Retention/Removal   187

11

Inference   201

12

Preface

I guess you could call this book the missing manual for **Elevation of Privilege (EoP)**. It explains how to play the game and is a reference guide, with example threats for the cards in the EoP card game, enabling threat modeling beginners to better understand what they might look for in their design.

Using *Elevation of Privilege* cards with the example threats in this book, you will discover the threats that could affect your software design through gameplay.

Each chapter covers a suit in the *Elevation of Privilege* card deck (a threat category), and for each card, example threats, references, and suggested mitigations are given. You'll cover the STRIDE with Privacy deck and the TRIM extension pack. **STRIDE** is a methodology for threat modeling and stands for **Spoofing, Tampering, Repudiation, Information Disclosure and Elevation of Privilege**, while **TRIM** is a framework for privacy and stands for **Transfer, Retention/Removal, Inference, and Minimization**. You'll learn what these terms refer to and how they should be applied.

You'll be able to recognize threats, understand privacy regulations, use the in-line references provided in the chapters to find out more, and learn what techniques you can use to protect against these threats and minimize your risk. Each threat is not necessarily one you will find in your design but is included to help you understand and consider what similar threats there could be in your design.

Who this book is for

The book is for the different participants involved in threat modeling:

- Security professionals and privacy engineers can use the book as a reference and support material when either facilitating or participating in threat modeling with teams.

- Software engineers and software architects can use the book when threat-modeling their software design. It supports them with concrete examples of threats so that they will then know what to look for and how it should be mitigated in their design, and they'll be in a position to create better threat models and, above all, more secure software.

- Product managers who need to threat-model their product with the team and support them by giving additional context they may have can use the book.

- Students and engineers wanting to move into the application security space as a career choice can use the book to support them in their studies.

What this book covers

Chapter 1, *Game Play*, explores how the game is played, who should play, what you'll need, and resources that you may find useful (in addition to this book, obviously).

Chapter 2, *Spoofing*, covers example spoofing threats, suggested mitigations for each of those threats, as well as references, where you can get additional background on a threat and its potential mitigations.

Chapter 3, *Tampering*, discusses example tampering threats, suggests mitigations for each of those threats, as well as references, where you can get additional background on a threat and its potential mitigations. (You should start to see a theme here.)

Chapter 4, *Repudiation*, dives into example repudiation threats, suggests mitigations for each of those threats, as well as references, where you can get additional background on a threat and its potential mitigations.

Chapter 5, *Information Disclosure*, examines example information disclosure threats, suggests mitigations for each of those threats, as well as references, where you can get additional background on a threat and its potential mitigations.

Chapter 6, *Denial of Service*, explores example privacy threats from elevation of privilege with privacy, suggests mitigations for each of those threats, as well as references, where you can get additional background on a threat and its potential mitigations.

Chapter 7, *Elevation of Privilege*, covers example elevations of privilege threats from the Elevation of Privilege suit, suggests mitigations for each of those threats, as well as references, where you can get additional background on a threat and its potential mitigations.

Chapter 8, *Privacy*, discusses example denial-of-service threats, suggests mitigations for each of those threats, as well as references, where you can get additional background on a threat and its potential mitigations.

Chapter 9, *Transfer*, dives into example transfer threats from the TRIM extension, suggests mitigations for each of those threats, as well as references, where you can get additional background on a threat and its potential mitigations.

Chapter 10, *Retention/Removal*, examines example retention/removal threats from the TRIM extension, suggests mitigations for each of those threats, as well as references, where you can get additional background for a threat and its potential mitigations.

Chapter 11, *Inference*, explores example inference threats from the TRIM extension, suggests mitigations for each of those threats, as well as references, where you can get additional background on a threat and its potential mitigations.

Chapter 12, *Minimization*, covers example minimization threats from the TRIM extension, suggests mitigations for each of those threats, as well as references, where you can get additional background on a threat and its potential mitigations.

Glossary offers a glossary of terms.

Appendix offers references for further reading.

To get the most out of this book

You'll need a solid understanding of the system or feature you are designing.

You'll need to have an architecture diagram of the system.

You should have an open mind and shouldn't see finding threats as a failure but as a learning opportunity to improve the security of your products.

Get in touch

Feedback from our readers is always welcome.

General feedback: If you have questions about any aspect of this book, email us at customercare@ packtpub.com and mention the book title in the subject of your message.

Errata: Although we have taken every care to ensure the accuracy of our content, mistakes do happen. If you have found a mistake in this book, we would be grateful if you would report this to us. Please visit www.packtpub.com/support/errata and fill in the form.

Piracy: If you come across any illegal copies of our works in any form on the internet, we would be grateful if you would provide us with the location address or website name. Please contact us at copyright@packtpub.com with a link to the material.

If you are interested in becoming an author: If there is a topic that you have expertise in and you are interested in either writing or contributing to a book, please visit authors.packtpub.com.

Share Your Thoughts

Once you've read *Threat Modeling Gameplay with EoP*, we'd love to hear your thoughts! Scan the QR code below to go straight to the Amazon review page for this book and share your feedback.

https://packt.link/r/1804618977

Your review is important to us and the tech community and will help us make sure we're delivering excellent quality content.

Download a free PDF copy of this book

Thanks for purchasing this book!

Do you like to read on the go but are unable to carry your print books everywhere?

Is your eBook purchase not compatible with the device of your choice?

Don't worry, now with every Packt book you get a DRM-free PDF version of that book at no cost.

Read anywhere, any place, on any device. Search, copy, and paste code from your favorite technical books directly into your application.

The perks don't stop there, you can get exclusive access to discounts, newsletters, and great free content in your inbox daily

Follow these simple steps to get the benefits:

1. Scan the QR code or visit the link below

https://packt.link/free-ebook/978-1-80461-897-4

2. Submit your proof of purchase

3. That's it! We'll send your free PDF and other benefits to your email directly

1
Game Play

In this chapter, I'm going to walk you through what you need to play **Elevation of Privilege (EoP)** to threat model your software design. We are going to talk about how the participants should be selected to get the best results from **threat modeling** and why participants should have different roles in the project. Last but not least, we will see how to play the game and understand what's the end goal of playing the game – finding out as many threats as possible. However, before we get started with all these, I would like to begin with a couple of words on what threat modeling is, as well as when you should threat model and why.

Threat modeling is a process to identify threats to and design flaws in the system you are designing. A **threat** is something that could go wrong in the system you are designing; it may be open to attack, it may be subject to some failure, or it may be open to human error. A **mitigation** is a safeguard or protection you can put in place to protect against a threat or at least reduce the risk a threat poses. So, when we threat model, we are looking for what could go wrong, how we can improve the system to stop that from happening, and finally, deciding whether we're happy that even if the worst happened, it wouldn't be all that bad because we've done a pretty good job.

When should we start? You should be able to begin threat modeling from the moment you are able to draw what your system will do and what parts it is made up of. Threat modeling is not a one-off exercise; it should be performed continually as your system evolves and it should be performed during the design phase of each version, and if the design changes during development, the process should be repeated to reflect those changes. Now, let's look at why it should be performed so early in the **software development life cycle (SDLC)**.

When you build a house, it's built on foundations, and it could be extremely complicated if you need to change those foundations halfway through construction. Design flaws are usually very difficult and costly to remediate once a project is underway.

Implementation flaws, on the other hand, are not necessarily difficult to fix after the fact. Using the housing analogy again, fixing an error in the foundations may mean tearing down parts of a construction and starting again from the foundations, whereas using a faulty or weak lock in a door is simple to fix because doors are designed to support standard lock fittings, you can just change the component.

So, we can conclude that it is always a wise choice to threat model early as it's an upfront investment that pays dividends.

Threat modeling can be used as a process for finding or eliciting security flaws in the design of a software system, although you could threat model any system. **EoP** is a category of threat and it is from this that the EoP card game for threat modeling takes its name. The EoP game was invented to facilitate threat modeling in teams as it prompts the participants with types of threats too.

As such, we will be covering the following main topics in the chapter:

- What you'll need to play the EoP game
- Who should participate?
- How to play EoP

By the end of the chapter, you will be familiar with the EoP card game, you will know where you can find useful resources to facilitate threat modeling with the game both remotely and in a single location, and you'll know who to invite.

What you'll need to play the EoP game

To get started, you're going to need a couple of things, depending on how you intend to play the game. Firstly, you are going to need a detailed architecture diagram showing the data flows and preferably the **trust boundaries**.

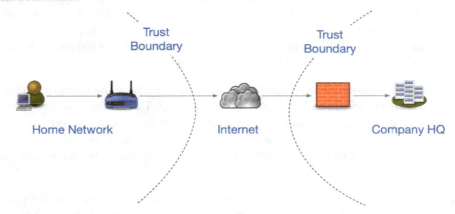

Figure 1.1: Diagram showing data flows and trust boundaries

What are the trust boundaries? They are the boundaries where data passes from one level of trust to another, for example, user input, which is untrusted data and data that has then been sanitized (had any invalid characters or commands removed), or data coming from the internet through the firewall and onto your network. In both cases, the second example is something you should be more willing to trust.

If you're going to be playing remotely, read the next section.

Having the cards either digitally or physically is going to be a help, so reading the section entitled *The cards* will point you to where you can download them digitally or purchase them online.

Remote threat modeling

If you're doing remote threat modeling exercises and you have a Miro account, you might find my **Threat Modeling with EoP** Miro template handy: `https://miro.com/miroverse/threat-modeling-with-eop/`.

The board contains instructions on how to get set up and a working example showing how the Miro board was intended to be used.

To deal with the cards for the remote exercise, **Agile Stationery** has kindly created a card-dealing web application:

`https://croupier.agilestationery.co.uk/`

Here, you can download *TNG Technology Consulting GmbH's* online multiplayer version of the threat modeling card games that you can host on-premises, such as **EoP**, **OWASP Cornucopia**, and **Cumulus**:

`https://github.com/tng/elevation-of-privilege`

The cards

The following resources are where you can get your hands on a copy of the EoP cards or those of one of its extensions required to play the game, either virtually or physically:

- Here's **Adam Shostack's GitHub repository** for EoP where you can download the cards: `https://github.com/adamshostack/eop`

- **Mark Vinkovits Privacy Extension** can be found here: `https://logmeincdn.azureedge.net/legal/gdpr-v2/eop-cards-ready-to-print.pdf`

- **F-Secure Corporation's Elevation of Privacy (T.R.I.M.) Extension** can be found here: `https://github.com/WithSecureOpenSource/elevation-of-privacye`

- You can buy physical copies of the cards from Agile Stationery as well: `https://agilestationery.com/collections/cybersecurity-games`

Alternative games

Two other threat modeling games that are quite similar to EoP in how you use them are Cornucopia from OWASP and Cumulus from TNG Technology. Many of the examples from this book will be applicable to cards in these games. Cornucopia is specifically designed for e-commerce applications and there are more threat categories, however, it doesn't map directly to STRIDE (which stands for the following threat categories: spoofing, tampering, repudiation, information disclosure, and EoP) if you have chosen to use this methodology. Cumulus, as the name suggests, is aimed at threat modeling cloud solutions. You can download these two games at the following links:

- **OWASP Cornucopia**: `https://owasp.org/www-project-cornucopia/#div-cards`
- **TNG Technology Consulting GmbH's Cumulus Cloud Threat Modeling Cards**: `https://github.com/TNG/cumulus`

Now that we have the resources we need to play the game, let's see who you should invite to play this game

Who should participate?

Preferably, you want between four and six players, covering different roles in the project and not necessarily technical roles. For example, you should include the software architect, a frontend/UI engineer if there is a UI component to the system, a backend engineer, a quality engineer, someone from the product team, and perhaps someone from compliance with knowledge of your privacy policies. The reason you want people from these different roles is to have a broader context. The product team is usually customer facing and so will be able to add context from that side of things; compliance will know what customers have signed up for, and what regulations and certifications the company needs to maintain, which will give additional context. People in different roles usually think differently because there is a certain amount of neurodiversity, so something one person misses others might spot.

You might find that people from product and compliance don't believe they will be useful because they may not feel they have the technical background. An analogy I like to use to make them more comfortable and feel more at ease is that you don't have to be a locksmith to know that if your key breaks in your front door lock, there is nobody home, and you've not got a key for another door, then you have a problem.

Now that we have our resources and we've invited the team members, we need to play the game. Let's see how the game is played.

How to play EoP

It's like any other card game, in so far that you win hands by playing the highest card. You have different suits; the cards have values and the *aces* are high cards. You win the hand by playing the highest card either of the same suit or by playing a trump card. With some variants, the cards go beyond ace as you will see in future chapters.

The difference is the objective, which is to find as many threats as possible, and if helping one another means you achieve that objective, then even better. It might seem less competitive that way, but later, you will see there are ways around that.

If you think of each hand as a battle and the game as a war, then what I am about to tell you will make sense. During each hand, you get points for finding threats, and those points, although won't win the hand, will accumulate and may mean you win the game.

Preparation

To play the game, you should deal the cards to each player until all the cards have been dealt. Depending on which variant you are playing, you will have between 6 and 11 suits. You can remove suits to reduce the time required / scope of the exercise if playing remotely. You can do this using the **Croupier** app and then distribute the cards to the players over chat or email, or, if you are all together, you can deal from a deck.

Aim

As the aim is to find as many threats as possible, players should avoid thinking about mitigations. This means they shouldn't think, "We've already protected against that type of threat so it's not valid anymore." Instead, they should think, "Aren't we clever spotting that threat and documenting both the threat and the protection that was put in place?"

Take the example of **Transport Layer Security (TLS)** or **Hypertext Transfer Protocol Secure (HTTPS)** for encryption in transit (sending data securely); not using it is a threat, using it means you have mitigated that threat (put a safeguard in place), and, as such, you should document this as part of your model. So, players should try and think where something can happen and then determine whether there is protection in place, document it, and, if not, propose one.

Why document something that has already been considered? So that if, at some point in the future, you are the victim of a threat actor and your company is held accountable, you can show that you did your due diligence and tried to protect your customers from as many threats as possible to the best of your ability.

To start

The player with the *3 of tampering* starts the game. They should read out the card they are playing for the benefit of the other players. They should look at their architecture diagram and try to recognize where the threat described on the card can occur. In the case of the *3 of tampering*, "*An attacker can take advantage of your custom key exchange or integrity control, which you built instead of using standard crypto,*" they should look for anywhere that cryptography or hashing is being performed in your architecture. If you are using standard crypto or hashing, then the threat still exists, and you can add what you are using as the mitigation of this threat.

If the player cannot find the threat or is unsure how the threat might occur, other players can help or make suggestions. They can also make suggestions of other places where the threat might occur. As a variant of the standard game, you could use this to assign extra points or even to steal points from other players. This can keep going until all other places where the threat can occur have been exhausted.

Don't forget!

It's a card game, so it should be fun as well. Like any other card game, the player who plays the highest card in the suit chosen at the start of the hand wins the hand.

There is a catch though; *Elevation of Privilege* cards are trump cards and if a player doesn't have a card in the suit you are playing, they can play a trump card. Playing a trump card doesn't guarantee you'll win the hand either, though, as someone might play a higher trump.

Points

Winning the hand gets you a point. As the point of the game is to find threats, finding a threat also gets you a point. The way I play it, finding multiple threats in a hand can get you a bonus point. This makes it possible to get a maximum of three points for your card in a hand, one if you win the hand, one for finding at least one threat, and one if you find any additional threats. You can, however, get extra points for finding occurrences of a threat for the card of an opponent.

So, how many points can you make in a hand? Let's see:

- One for winning the hand
- One for finding your threat
- One for finding more occurrences of your threat
- n (players − 1) points if you find a threat for each card that your opponents play

  This means that if there are six players, you could get eight points in a hand.

You might consider giving points for suggesting mitigations for any new threats found, but you can decide as a team what works best for you.

Who goes next?

If you're playing in a room, it could be the person next clockwise or anticlockwise around the table; if you're playing online, it could be whoever was next when the names were put into the Croupier app. It doesn't really matter; just make a note of the order for future hands.

When one hand finishes, the winner of the hand (not who has the most points) gets to choose what suit comes next and they open the hand playing the first card. The player after them will be whoever followed them in the last hand.

Keep going until you've run out of suits or cards in your hand, whichever you prefer.

While playing, you should be making a note of the threats found on the scorepad, potentially creating tickets for those threats and proposals for mitigating them. If you're playing remotely, this can be done by adding stickies to the collaboration board; I've used red ones for threats, green ones for mitigations, either already implemented or already in the design, and orange ones for mitigation proposals.

Who's won?

The customer, because the product is more secure.

Joking aside, whoever has accrued the most points during the game is the winner, just like any other game. What do you win? Well, that is entirely at the discretion of the team or your management. It could be kudos to your team, it could be a voucher, or it could be something else; I leave it up to your imagination.

Variations of gameplay

Some teams prefer to pick a suit and go through all the cards one by one discussing them as a team. This removes some of the gamification aspects but is still an effective way of threat modeling the architecture.

Other teams prefer individually adding threats where they believe they can occur simultaneously and then discussing each other's ideas once all the players are happy that they can't find any other threats. Again, it can be an effective means, but it removes some of the gamification aspects of threat modeling with the EoP game.

The group discussion can also be a very powerful means to spark ideas in others where something similar can occur. Some players favor this approach over another, perhaps because there is a very outspoken member of the team or because they are timid. If you are facilitating a threat modeling session, you should be aware of the team dynamic and you should try and help each player feel comfortable and able to give their input.

Obstacles

Initially, you may need to find teams that are willing to experiment and open to championing the approach with their colleagues. Product teams are often under pressure with tight deadlines; these deadlines are often driven by a need to sell new features. So, this is all the more reason to involve people in defining these deadlines because it will help them understand that the upfront investment could save time and effort later. Once they start to see the benefits, you will find the time is included in the planning.

Initially, there is a learning curve because teams will be learning the technique by doing, and engineers will undoubtedly complain that it takes time. As they improve, they will get faster, but initially, they will be threat modeling both the legacy and the new. However, soon they can concentrate on the new features.

Some will complain that there is repetition between projects; this is a problem relating to the documentation or processing of the models rather than the models themselves. I would recommend using what I call a **hybrid approach**. Using a tool that will allow you to draw your architecture from your existing models either through templating or as components will promote re-use. If the tool also offers some level of automated threat modeling, then even better. This will allow you to capture the basic threats or low-hanging fruit related to the standard components in your system, letting you concentrate on the proprietary technology in your system. It will also speed up the process.

Scaling your threat modeling program

Gamified threat modeling is a great way to train engineering teams to threat model; it will help them develop the skills needed and they will be able to self-serve. The security team should still be involved but more in a supervisory role, reviewing threat modeling reports or offering consultancy when teams feel they need support with a particularly security-sensitive project.

As teams mature, members from one team will be able to facilitate for members in another team, allowing for accelerated diffusion of the know-how within the organization.

Again, a hybrid approach would also allow for your program to scale because teams would be able to make use of existing models of components parts of their system.

Performance metrics and reporting

Most organizations will already have metrics around the number of escaped vulnerabilities or issues found during penetration/security testing. Over time, you should see a reduction in these.

If you record the threats found during modeling and create tickets for all the suggested mitigations, labeling them as coming from threat modeling, you should be able to track them. Recording the number of threats found, the effort in implementing the mitigations, the reduction in the number of escaped vulnerabilities reaching pen-testing, and the associated average cost of those escaped vulnerabilities, should allow you to demonstrate the value of the program in monetary terms.

Coming up

In the coming chapters, I will introduce the chapter with a brief explanation or definition of what the threats category name means. Then, I give examples for each of the threats described on the cards; some cards may have multiple examples. Each example is structured as follows:

2. of EoP Suit

The description of the type of threat from the card is as follows:

Threat	
	A description of the example threat
CAPEC	One or more CAPEC entries that you can lookup
ASVS	One or more ASVS entries you can lookup
CWE	One of more CWE entries you can lookup

Mitigations	
✓	• A potential mitigation • Another potential mitigation

As you can see, the title of an example (*2. of EoP Suit* in this case) is the card value followed by its suit or threat category as you might prefer to call it. This is followed by the card description as you would read it on the face of the physical cards.

Next in the *red* threat table, an example threat is described to guide you and help you understand how this threat might manifest itself in a real-world application.

Following the example are references coming from **Mitre** and **Open Worldwide Application Security Project (OWASP)**:

- Mitre **Common Attack Pattern Enumeration and Classification (CAPEC)**, which you can look up here: `https://capec.mitre.org/index.html`

- Mitre **Common Weakness Enumeration (CWE)**, which you can look up here: `https://cwe.mitre.org/index.html`

- OWASP **Application Security Verification Standard (ASVS)**, which you can look up here: `https://owasp.org/www-project-application-security-verification-standard/`

- You can also use my **CAPEC STRIDE Mapping Mind** maps that map the CAPEC entries to the **STRIDE** categories and it contains many threats you won't find on the cards that you may be able to use for the ACE cards: `https://www.ostering.com/blog/2022/03/07/capec-stride-mapping/`

CAPEC

CAPEC is a directory containing almost all known threats, created by Mitre with the following license: `https://capec.mitre.org/about/termsofuse.html`.

Each threat in the directory is classified and any associated threats, macro categories, or child categories are included along with a detailed description of the threat.

CWE

CWE is a directory containing an extensive list of software and hardware weaknesses that cause vulnerabilities, created by Mitre under the following license: `https://cwe.mitre.org/about/termsofuse.html`.

Each weakness in the directory is classified, and any related weaknesses, macro categories, or child categories are included, along with a detailed description of the weakness.

STRIDE

STRIDE is a framework for threat modeling and was invented at Microsoft by Praerit Garg and Loren Kohnfelder. The framework helps by giving you key threat types, which can help you reason where the software architecture might be susceptible. In EoP, these categories are used for the different suits in the card deck. CAPEC has its own classification and isn't classified according to STRIDE, so I created the mind maps to help you if you want to advance your threat modeling skills further.

Important note

There are three things you can do to protect yourself from a risk:

a. One is to mitigate the risk (you would use compensating controls here)

b. Another is to transfer the risk (insurance, terms and conditions, and contracts are all examples of this)

c. The last is to avoid the risk (don't do what it is that causes the risk, for example, skydiving has a risk of death if your parachute doesn't open; if you don't do skydiving, the risk of dying from skydiving doesn't exist)

Ignoring a threat is not something that will reduce your risk.

Next, the *green* table contains a list of potential mitigations or compensating controls that in some cases will reduce the risk of the threat, in others they may remove the risk of the threat entirely. You can use a combination of multiple mitigations to reduce the risk even further in some cases.

Summary

In this chapter, you've learned the following:

- What you'll need to perform the threat modeling session.

- Where you can get the cards to play the game.

- The different decks of cards you have available.

- Where you can find a number of additional resources to support you in threat modeling, including a Miro template for remote threat modeling, a web app for dealing the cards remotely, and an online version of the game that you can run on your local network.

- Who should participate in the threat modeling and why.

- How you play the game and what the aim of playing EoP is.

- Some variations of play that may work better for you as a team.

I've then given you a sneak peek at what's to come in the next chapters and where you can go and look up all those references you'll see going through the book.

In the following chapters, you will see examples for each of the cards in EoP and some of its extensions. Having read this chapter, you are now in a position to jump straight in and start threat modeling, and for each card being played, either by you or other players, you can look it up and see one or more examples to help you understand the threat and give you an indication of where it might occur in your architecture. Happy threat modeling!

2
Spoofing

Spoofing is when you make a fake or hoax of something, usually maliciously to trick the other party. Phishing is a type of spoofing where you might receive an email that appears to be from your bank but, in fact, it's tricking you into giving the bad actor some piece of data they need. Phishing (faking an email from an organization or person you may have dealings with) is a form of spoofing. Often, social engineering attacks are forms of spoofing, such as vishing, which is a telephone call from someone pretending to be from your bank to trick you into performing some action. There are also many other types of spoofing, as we will see as we explore each of the cards in the Spoofing suit.

In this chapter, we'll go through a series of example spoofing threats from the Spoofing suit in the Elevation of Privilege card game. We'll see a variety of references where you can read more about the threat and we'll also give you suggestions on how to mitigate the threat.

By the end of the chapter, you should be able to recognize where you are potentially introducing spoofing design flaws when designing your applications and be able to change your design accordingly, or at least find these issues when you threat model your design.

Figure 2.1: An attacker identity spoofing

2. of Spoofing

An attacker could squat on the random port or socket that the server normally uses.

Or the alternative text:

An attacker could take over the port or socket that the server normally uses.

Threat	
	Also known as scheme squatting. An attacker could bind to port 80, the port used for the **HyperText Transfer Protocol** (**HTTP**), and respond to requests instead of your app/service. If an attacker responds to your service calls instead of the service, they can feed your consumer fake data.
CAPEC *[1]*	CAPEC-616 - Establish Rogue Location CAPEC-505 - Scheme Squatting
ASVS *[2]*	N/A
CWE *[3]*	CWE-200 - Exposure of Sensitive Information to an Unauthorized Actor

Mitigations	
	• Use **Transport Layer Security** (**TLS**) for your service so that its client can verify the authenticity of the application. • Ensure certificates are signed by a trusted **Certificate Authority** (**CA**). • Ensure that certificates have not expired or been revoked. • Enable **Multi-Factor Authentication** (**MFA**). • Use reserved ports. • Restrict who can install applications. • Use an alert for configuration changes such as `Config` in AWS. • Overwrite configuration regularly using Puppet to ensure constancy.

3. of Spoofing

An attacker could try one credential after another and there's nothing to slow them down (online or offline).

Threat	
	Brute forcing a login form until the password matches. They might choose to try the username admin and then, using a dictionary of common passwords, try to guess the right password.
CAPEC	CAPEC-49 - Password Brute Forcing CAPEC-16 - Dictionary Based Password Attack CAPEC-565 - Password Spraying
ASVS	2.2.1 - Ensure you have protections in place against automated attacks and that they are tested
CWE	CWE-307 - Improper Restriction of Excessive Authentication Attempts

Mitigations	
	• Add increasingly large pauses between failed attempts (i.e., a back-off strategy). • Use Captchas. • Add a temporary lockout after *n* failed attempts. • Enable MFA.

4. of Spoofing

An attacker can anonymously connect because we expect authentication to be done at a higher level.

Threat	
	Addressing pages directly that you would normally reach via a login flow. If you know the address and the login was in the flow, it may be possible to bypass, thereby indicating a missing object-level access control.
CAPEC	CAPEC-87 - Forceful Browsing
ASVS	4.2.1 - Ensure object level access control is implemented correctly and authorizations are performed on every request
CWE	CWE-425 - Direct Request ('Forced Browsing')

Mitigations	
	• Ensure that you have object-level access control.
	• Ensure that you are always performing authorization checks.
	• Always implement *secure by default* and then relax permissions.

5. of Spoofing

An attacker can confuse a client because there are too many ways to identify a server.

Threat	
	Here are some examples of addressing the same host, which can become confusing: • `https://server1.mydomain.com` • `https://server1` • `https://192.168.1.10` • `https://0xC0.A8.1.A` • `https://030052000412` Not to mention **Domain Name Service** (DNS) aliases, IPv6 addresses, additional IPs, and so on.
CAPEC	CAPEC-4 - Using Alternative IP Address Encodings
ASVS	N/A
CWE	CWE-173 - Improper Handling of Alternate Encoding CWE-647 - Use of Non-Canonical URL Paths for Authorization Decisions

Mitigations	
	• Use standard naming conventions and fully qualify them • Use the canonical form

6. of Spoofing

An attacker can spoof a server because identifiers aren't stored on the client and checked for consistency on reconnection (that is, there's no key persistence).

Threat	
	You are not checking known hosts for Secure Shell (SSH) connections because you have the `StrictHostKeyChecking` configuration setting set to no, which allows an attacker to spoof the server because the terminal application will not inform you that the certificate presented during connection has changed.
CAPEC	CAPEC-195 - Principal Spoof
ASVS	9.2.1 - Ensure you're verifying TLS certificates and internal certificate authority certificates or self-signed certificates are also in your trust store
CWE	CWE-295 - Improper Certificate Validation

Mitigations	
	• Host verification should always be active where possible • Alert if new IDs are discovered

7. of Spoofing I

An attacker can connect to a server or peer over a link that isn't authenticated (and encrypted).

Threat	
	You have an internal API so that your applications can perform lookups of customer details (addresses, phone numbers, etc.) but, although only intended for internal use, this API doesn't require authentication and access from the internet hasn't been blocked. So, an attacker, having discovered the service, is now harvesting all your customer information.
CAPEC	CAPEC-36: Using Unpublished Interfaces or Functionality
ASVS	1.2.2 Ensure all your APIs that expose sensitive data are authenticated
CWE	CWE-306: Missing Authentication for Critical Function

Mitigations	
	• Always use encryption in transit where possible, and authenticate and authorize users
	• Adopt a zero-trust culture and encrypt local network traffic as well
	• Put your internal APIs behind the firewall
	• Segment your network and limit the incoming connections to only the hosts you expect
	• Make sure that any APIs that allow access to data are authenticated
	• Use an industry-standard proven authentication schema (method of authentication)

7. of Spoofing II

An attacker can connect to a server or peer over a link that isn't authenticated (and encrypted).

Threat	
	The attacker could use the unencrypted version of a protocol such as HTTP instead of HTTPS to connect to the machine.
CAPEC	CAPEC-94 - **Adversary in the Middle (AiTM)**
ASVS	1.9.1 - Ensure you're using TLS everywhere 9.1.1 - Ensure the TLS version can't be downgraded
CWE	CWE-319 - Cleartext Transmission of Sensitive Information

Mitigations
• Always use encryption in transit where possible, and authenticate and authorize users. • Adopt a zero-trust culture and encrypt local network traffic as well.

8. of Spoofing

An attacker could steal credentials stored on the server and reuse them (for example, a key is stored in a world-readable file).

Threat	
	You've stored a private key in a PEM file that isn't encrypted, and access isn't restricted.
CAPEC	CAPEC-191 - Read Sensitive Constants Within an Executable CAPEC-150 - Collect Data from Common Resource Locations
ASVS	2.10.4 - Ensure Keys and Secret material are stored securely and ensure that secrets aren't hardcoded in source files
CWE	CWE-257 - Storing Passwords in a Recoverable Format CWE-256 - Plaintext Storage of a Password CWE-522 - Insufficiently Protected Credentials CWE-798 - Use of Hard-coded Credentials

Mitigations	
	• Encrypt the key • Use strong ACLs • Use a PAM tool to manage keys

9. of Spoofing I

An attacker who gets a password can reuse it (use stronger authenticators).

Threat	
	An attacker might shoulder surf and read what you are typing or use a key logger attached to your computer to steal your password, which they can then reuse because you don't require additional factors (token, biometric, FIDO2).
CAPEC	CAPEC-560 - Use of Known Domain Credentials
ASVS	2.2.6 - Verify replay attack protections are in place and working correctly
CWE	CWE-308 - Use of Single-factor Authentication

Mitigations	
	• Your application should require MFA

9. of Spoofing II

An attacker who gets a password can reuse it (use stronger authenticators).

Threat	
	If the user has reused their password and a service provider has not been protecting their data correctly, their password may have been stolen, which they can then reuse because you don't require additional factors (token, biometric, FIDO2).
CAPEC	CAPEC-560 - Use of Known Domain Credentials
ASVS	2.2.6 - Ensure replay attack protections are in place and working correctly 2.2.7 - Ensure user-in-the-loop with automation protection controls
CWE	CWE-308 - Use of Single-Factor Authentication

Mitigations	
	• Your application should require MFA

10. of Spoofing

An attacker can choose to use weaker or no authentication.

Threat	
	An attacker can remove the signature algorithm from a JWT exploiting a "None Algorithm" vulnerability and allowing them to change a token and gain access to your systems.
CAPEC	CAPEC-620 - Drop Encryption Level
ASVS	3.5.3 - Ensure Null Cipher Checks are implemented
CWE	CWE-757 - Selection of Less-Secure Algorithm During Negotiation ('Algorithm Downgrade') CWE-345 - Insufficient Verification of Data Authenticity

Mitigations	
	• Ensure that you check that the signature algorithm is the one you expect and that you are also verifying that the signature is valid before accepting any of the claims in the token

Jack of Spoofing

An attacker could steal credentials stored on the client and reuse them.

Threat	
	An attacker could steal your session cookies or use something such as a key logger to capture your credentials when logging in.
CAPEC	CAPEC-568 - Capture Credentials via Keylogger CAPEC-31 - Accessing/Intercepting/Modifying HTTP Cookies
ASVS	3.4 and 3.5 - Ensure Cookies are secured properly and only accessible from the source host 2.2.6 - Ensure replay attack protections are in place and working correctly 2.2.7 - Ensure user-in-the-loop with automation protection controls 3.4.1, 3.4.2, 3.4.3, 3.4.4, 3.5.1 - Ensure Cookies are secured properly and only accessible from the source host 2.2.6 - Ensure replay attack protections are in place and working correctly 2.2.7 - Ensure user-in-the-loop with automation protection controls
CWE	CWE-79 - Improper Neutralization of Input During Web Page Generation ('Cross-site Scripting') CWE-539 - Use of Persistent Cookies Containing Sensitive Information

Mitigations	
	<ul><li>Require MFA</li><li>Detect multiple instances of the same session and send a notification</li><li>Fingerprint the user's machine and associate their characteristics with the session</li><li>Geo-fencing of users</li><li>Use malware endpoint protection to block any attempt to deploy the key logger</li><li>Add the `Secure`, `HttpOnly`, `SameSite`, `Domain`, and `Path` attributes to the cookie</li><li>Perform necessary SAST and DAST scans to test the controls you have in place</li></ul>

Queen of Spoofing I

An attacker could go after the way credentials are updated or recovered (account recovery doesn't require disclosing the old password).

Threat	
	If the reset asks questions such as what your mother's maiden name is, or whether you or your parents are famous, this information may be in the public domain, so they can use this to reset your password to what they want. Alternatively, if your family tree is online, then it is equally likely that an attacker could find this information.
CAPEC	CAPEC-50 - Password Recovery Exploitation
ASVS	2.1.6 - Ensure both the new and current password are required to change password
	2.2.3 - Verify notifications sent for password changes
	2.5.2 - Verify password hints or security questions aren't used
	2.5.6 - Verify password recovery requires an additional authentication factor/channel
CWE	CWE-640 - Weak Password Recovery Mechanism for Forgotten Password
	CWE-620 - Unverified Password Change

Mitigations	
	• Require MFA
	• Prevent/disable usage of security questions
	• Use fake values for security question answers

Queen of Spoofing II

An attacker could go after the way credentials are updated or recovered (account recovery doesn't require disclosing the old password).

Threat	
	Unless an email is sent using Transport Layer Security (TLS), an attacker could request a reset password and intercept the email containing a link to reset the password.
CAPEC	CAPEC-50 - Password Recovery Exploitation
ASVS	2.1.6 - Ensure both the new and current password are required to change the password
	2.2.3 - Ensure notifications are sent for password changes
	2.7.2 - Ensure verifiers have short TTLs (time to live)
	2.7.3 - Ensure verifiers are Single Use
	2.7.4 - Ensure verifiers communicated over a secure channel
CWE	CWE-640 - Weak Password Recovery Mechanism for Forgotten Password
	CWE-620 - Unverified Password Change

Mitigations	
	• Use TLS for sending reset emails
	• Make links include a random value
	• Make links expire after a single use
	• Make links with a short TTL

King of Spoofing I

Your system ships with a default admin password and doesn't force a change.

Threat	
	When you buy a router, it has the admin password on the back. Some vendors use the same one for everyone and detail it in the user manual.
CAPEC	CAPEC-70 - Try Common or Default Usernames and Passwords
ASVS	2.5.4 - Ensure shared or default accounts have been removed 2.3.1 - Ensure forced change of password on first login
CWE	CWE-1392 - Use of Default Credentials CWE-1393 - Use of Default Password

Mitigations	
	• Change the password immediately and rotate it (change it on a regular schedule) regularly

King of Spoofing II

Your system ships with a default admin password and doesn't force a change.

Threat	
	New users are created with a default password that is always the same and they aren't obliged to change it on first login. This results in many new users using the same password if not required to be changed on a regular schedule, potentially indefinitely.
CAPEC	CAPEC-70 - Try Common or Default Usernames and Passwords
ASVS	2.5.4 - Ensure shared or default accounts have been removed 2.3.1 - Ensure forced change of password on first login
CWE	CWE-1392 - Use of Default Credentials CWE-1393 - Use of Default Password

Mitigations	
	• Use a random password for each new user • Require users to change their password on the first login • Require users to change their passwords regularly

Ace of Spoofing I

You've invented a new spoofing attack.

Threat	
	If the system has end users, this could be a phishing attack that takes advantage of the user already being logged in, such as **Cross-Site Request Forgery (CSRF)**.
CAPEC	CAPEC-62 - Cross-Site Request Forgery
ASVS	4.2.2 - Ensure you are protecting against Cross-Site Request Forgery (CSRF)
CWE	CWE-352 - Cross-Site Request Forgery (CSRF)

Mitigations
• Use CSRF tokens hidden in web forms that may be either for the duration of a session or, better, still valid since the last request. In this way, the attacker won't have them in their possession to be able to trick you into performing some action by clicking on a link.

Ace of Spoofing II

You've invented a new spoofing attack.

Threat	
	Another type of spoofing not in the other cards is DNS spoofing to trick the user or service into talking to a host on a different IP.
CAPEC	CAPEC-142 - DNS Cache Poisoning CAPEC-598 - DNS Spoofing
ASVS	10.3.3 - Ensure good DNS hygiene practices
CWE	CWE-350 - Reliance on Reverse DNS Resolution for a Security-Critical Action CWE-290 - Authentication Bypass by Spoofing CWE-295 - Improper Certificate Validation CWE-923 - Improper Restriction of Communication Channel to Intended Endpoints

Mitigations	
	• Using DNSSEC, the DNS records are signed, which makes it more difficult for them to be spoofed

Ace of Spoofing III

You've invented a new spoofing attack.

Threat	
	An attacker may try to trick customer support into giving them the password or sending them a link to reset the password; this is a social engineering attack called **vishing** (short for **voice phishing**). The attacker could also make use of data gathered from social media and the dark web to respond to questions from the operator. This information gathering is called **Open Source Intelligence (OSINT)**.
CAPEC	CAPEC-656: Voice Phishing CAPEC-98: Phishing
ASVS	2.2.4 Ensure there is some protection against impersonation, such as asking the user for a one-time password or some other authentication factor
CWE	CWE does not currently cover social engineering in the way it is presented by CAPEC.

Mitigations	
	• Security awareness training programs for employees • Phishing simulation tools validate an employee's skills in recognizing an attack and can be used to determine the risk they pose to the organization For both mitigations, many security vendors have a solution such as Mimecast, KnowBe4, or Cofense

Ace of Spoofing IV

You've invented a new spoofing attack.

Threat	
	To trick an employee who works in the finance department, an attacker may send a message via SMS or a messaging service such as WhatsApp claiming to be a company executive or someone of importance, asking them to make an urgent bank transfer because they are offsite and it needs to be done immediately. Because of the urgency of the message and the status of the person being impersonated, the employee may feel compelled to make the transfer. These methods are often used to manipulate/ social engineer the victim. This type of attack is known as smishing.
CAPEC	CAPEC-164: Mobile Phishing
ASVS	N/A
CWE	CWE does not currently cover social engineering in the way it is presented by CAPEC.

Mitigations	
	• Security awareness training programs for employees.
	• Phishing simulation tools validate an employee's skills in recognizing an attack and can be used to determine the risk they pose to the organization
	• Implement company policies and procedures that define the protocols that should be used for both making and accepting urgent requests of this nature

E. of Spoofing

We cannot tell which of our admins edited personal data, as admin accounts are shared.

Threat	
	Your administrators are using the same login credentials, perhaps to save on paying for extra licenses. This isn't a good practice because it gives plausible deniability in the event that they perform some accidental or nefarious action.
CAPEC	CAPEC-560 - Use of Known Domain Credentials CAPEC-653 - Use of Known Operating System Credentials
ASVS	2.5.4 - Ensure shared or default accounts have been removed 2.10.1 - Ensure services are not authenticating with shared accounts
CWE	N/A

Mitigations	
	• Ensure that each administrator has their own account • Credentials should not be shared • Actions should be logged

Summary

You've now covered the threat types described on the cards from the first suit, Spoofing, in the Elevation of Privilege card deck, with the addition of one card from the T.R.I.M. extension for the game. These threats detailed flaws relating to both encryption in transit and encryption at rest, as well as flaws relating to authentication and password security and flaws relating to addressing.

Having a greater awareness of the threats in this category and how to mitigate them should help you to design more secure software and enable you to recognize these and similar flaws when threat modeling.

In the next chapter, we will cover the second S.T.R.I.D.E. category of threats, Tampering.

References

- [1] **Mitre Common Attack Pattern Enumeration and Classification** (**CAPEC**) which you can look up here: `https://capec.mitre.org/index.html`

- [2] **OWASP Application Security Verification Standard** (**ASVS**) which you can look up here: `https://owasp.org/www-project-application-security-verification-standard/`

- [3] **Mitre Common Weakness Enumeration** (**CWE**) which you can look up here: `https://cwe.mitre.org/index.html`

3
Tampering

Tampering is the act of interfering with or modifying something with the intent to cause damage. This could be an act of vandalism; it could be disabling a security system to gain access, as shown in *Figure 3.1*, or it could be modifying an identity document to gain access to a facility. Tampering comes in many shapes and sizes.

In this chapter, we will cover the different tampering threats described on the cards available in the Tampering deck in the Elevation of Privilege card game. Tampering could affect your data both at rest or in transit, the software systems you use, and the software or hardware you develop. You'll see that these threats are often caused by design flaws relating to access control, missing integrity checks, and missing encryption both at rest and in transit.

By the end of this chapter, you should be able to identify when these flaws are present in a design and potential strategies you can use to mitigate the threat or at least reduce the risk it poses to a manageable level.

Figure 3.1: Tampering with surveillance

2. of Tampering (2022 deck) I

An attacker can modify your build system and produce signed builds of your software:

Threat	
	Your credentials have been stolen and an attacker uses them to log in to the **continuous integration/continuous delivery (CI/CD)** platform. They modify the workflow so that it includes malware in the build artifact (JAR/executable/install files).
CAPEC	CAPEC-678: System Build Data Maliciously Altered
	CAPEC-445: Malicious Logic Insertion into Product Software via Configuration Management Manipulation
	CAPEC-670: Software Development Tools Maliciously Altered
	CAPEC-446: Malicious Logic Insertion into Product via Inclusion of Third-Party Component
	CAPEC-511: Infiltration of Software Development Environment
	CAPEC-538: Open-Source Library Manipulation
	CAPEC-186: Malicious Software Update
	CAPEC-442: Infected Software
ASVS	1.1.1: Implement a **secure software development life cycle (SSDLC)**
	1.10.1: Verify the security and accountability of source code control
	1.14.3: Ensure vulnerable second and third-party components are not being used
CWE	CWE-733: Compiler Optimization Removal or Modification of Security-critical Code

Mitigations
• Require VPN access to reach the CI/CD platform • Enable MFA to the CI/CD platform • Verify signatures during build • Geo-fence users who can modify build jobs

2. of Tampering (2022 deck) II

An attacker can modify your build system and produce signed builds of your software:

Threat	
	Your laptop has been stolen and an attacker uses it to commit malicious code to a project in your code repository.
CAPEC	CAPEC-445: Malicious Logic Insertion into Product Software via Configuration Management Manipulation

CAPEC-670: Software Development Tools Maliciously Altered

CAPEC-511: Infiltration of Software Development Environment |
| ASVS | 1.10.1: Verify the security and accountability of source code control

10.2.5: Ensure you are checking for malicious code in your code base |
| CWE | CWE-506: Embedded Malicious Code |

Mitigations	
	• Require VPN access to reach the code repository

• Enable MFA to the code repository

• Sign commits to the code repository

• Verify signatures during build

• Geo-fence users who can commit to the repository

• Require at least one peer review for all merge/pull requests |

2. of Tampering (2022 deck) III

An attacker can modify your build system and produce signed builds of your software:

Threat	
	You're using an open source third-party library and an attacker modifies the library and uploads it to the public repository, which, when downloaded during the build, gets integrated into your product.
CAPEC	CAPEC-446: Malicious Logic Insertion into Product via Inclusion of Third-Party Component CAPEC-538: Open-Source Library Manipulation CAPEC-186: Malicious Software Update CAPEC-442: Infected Software
ASVS	1.14.3: Ensure vulnerable second and third-party components are not being used
CWE	CWE-507: Trojan Horse

Mitigations	
	• Verify the signatures of third-party components • Use an internal artifact repository • Quarantine artifacts until they're scanned for known vulnerabilities • Quarantine artifacts until they're scanned for malware • Test the artifacts in a sandbox

2. of Tampering (2022 deck) IV

An attacker can modify your build system and produce signed builds of your software:

Threat	
	You're using an open source third-party library and an attacker modifies the library and uploads it to the public repository, which, when downloaded, compromises your build environment.
CAPEC	CAPEC-678: System Build Data Maliciously Altered
	CAPEC-445: Malicious Logic Insertion into Product Software via Configuration Management Manipulation
	CAPEC-670: Software Development Tools Maliciously Altered
	CAPEC-511: Infiltration of Software Development Environment
	CAPEC-538: Open-Source Library Manipulation
ASVS	10.2.1: Ensure the source code doesn't contain a phone home mechanism
	10.2.3: Ensure the source code doesn't contain malicious payloads or backdoors
CWE	CWE-507: Trojan Horse

Mitigations	
	• Verify the signatures of third-party components • Use an internal artifact repository • Quarantine artifacts until they're scanned for known vulnerabilities • Quarantine artifacts until they're scanned for malware • Test the artifacts in a sandbox

3. of Tampering

An attacker can take advantage of your custom key exchange or integrity control, which you built instead of using standard crypto:

Threat	
	You've written a hashing algorithm or an encryption algorithm instead of using a known proven algorithm.
CAPEC	CAPEC-20: Encryption Brute Forcing
ASVS	6.2.2: Ensure standard-recognized and tested cryptographic algorithms are being used
CWE	CWE-1240: Use of a Cryptographic Primitive with a Risky Implementation

Mitigations	
	• Use the best standard algorithm you can at the time for the type of data you are encrypting – that is, stream or file data.
	• Make your system cryptographically agile so that you can migrate to a new algorithm without too much difficulty if the algorithm you're using breaks.
	• Building crypto is hard and even the experts get it wrong. Unless your algorithm has been peer-reviewed by the global community and you're a specialist in cryptographic algorithms and mathematics, it is not generally a good idea to roll your own.

4. of Tampering I

Your code makes access control decisions all over the place, rather than with a security kernel:

Threat	
	You haven't centralized your access control and in different areas of your application, the checks against roles are hard-coded. So, if you change the roles later, you have to remember to make the changes in all the different parts of the system.
CAPEC	CAPEC-180: Exploiting Incorrectly Configured Access Control Security Levels CAPEC-1: Accessing Functionality Not Properly Constrained by **Access Control Lists (ACLs)**
ASVS	1.4.4: Ensure the use of a security kernel to manage access control
CWE	CWE-280: Improper Handling of Insufficient Permissions or Privileges CWE-266: Incorrect Privilege Assignment

Mitigations	
	• Centralize the functions that check if you are in the right role to perform a particular action. That way, you can change the name of a role retaining the same ID without having to modify the code in multiple points, risking that something is missed.

4. of Tampering II

Your code makes access control decisions all over the place, rather than with a security kernel:

Threat	
	You're checking roles and making decisions based on a fixed set of roles but haven't considered that new roles may be added and your code doesn't have a secure default behavior. As these checks are not centralized, you must go and touch multiple parts of the code base and forget to modify all the places where these checks have been implemented.
CAPEC	CAPEC-180: Exploiting Incorrectly Configured Access Control Security Levels CAPEC-1: Accessing Functionality Not Properly Constrained by ACLs
ASVS	1.4.4: Ensure the use of a security kernel to manage access control
CWE	CWE-280: Improper Handling of Insufficient Permissions or Privileges CWE-266: Incorrect Privilege Assignment CWE-842: Placement of User into Incorrect Group

Mitigations	
	• Checks should be central • Checks should include the subject, object, and action, depending on the access control being applied (MAC, DAC, RBAC, or ABAC)

5. of Tampering I

An attacker can replay data without detection because your code doesn't provide timestamps or sequence numbers:

Threat	
	You have an offer with a discount code that, during your checkout process, a customer applies more than once and you aren't verifying if that code has been used by that customer already.
CAPEC	CAPEC-77: Manipulating User-Controlled Variables
ASVS	11.1.3: Ensure that user limits and restrictions are in place for transactions
CWE	CWE-837: Improper Enforcement of a Single, Unique Action

Mitigations	
	• When a discount code is issued, it should only be used once for a given order. Store the code and the order ID in a record so that you can check if the discount has already been applied.

5. of Tampering II

An attacker can replay data without detection because your code doesn't provide timestamps or sequence numbers:

Threat	
	You are using SAML and your authn response doesn't contain a sequence number or the sequence number isn't being considered. An attacker can steal the token and can then reuse it to perform actions until it expires.
CAPEC	CAPEC-21: Exploitation of Trusted Identifiers
ASVS	3.5.3: Ensure tokens are signed and encrypted 3.6.1: Ensure users must re-authenticate within a defined timeframe 3.6.2: Ensure relying parties are aware when a user last authenticated
CWE	CWE-613: Insufficient Session Expiration CWE-306: Missing Authentication for Critical Function CWE-287: Improper Authentication

Mitigations	
	• Add a sequence number to the tokens • Store used sequence numbers for each user • Verify against stored sequence numbers before accepting a token

6. of Tampering

An attacker can write to a data store your code relies on:

Threat	
	Your data store, be it a database, filesystem, or object storage, doesn't require authentication.
CAPEC	CAPEC-180: Exploiting Incorrectly Configured Access Control Security Levels CAPEC-470: Expanding Control over the Operating System from the Database CAPEC-592: Stored XSS
ASVS	1.2.2: Verify access to the data layer is authenticated and the principle of least privilege is being used
CWE	CWE-921: Storage of Sensitive Data in a Mechanism without Access Control

Mitigations	
	• Require authentication and authorization • Require TLS • Restrict access to predefined IPs • Remove direct exposure to the internet

7. of Tampering I

An attacker can bypass permissions because you don't make names canonical before checking access permissions:

Threat	
	You're accepting user-supplied paths without getting the canonical form before making access control decisions.
	A user can supply a relative path, allowing them to access other areas of the filesystem. For example, by supplying the `../../etc/` path, the user can navigate back up two levels in the tree to the root and then into the `/etc` folder: `/webapp/uploads/../../etc/`.
	In this case, they may be able to overwrite files in the `/etc` folder.
CAPEC	CAPEC-126: Path Traversal
	CAPEC-139: Relative Path Traversal
ASVS	12.3.1: Ensure that a URL API is used to protect against path traversal attacks
CWE	CWE-73: External Control of File Name or Path
	CWE-41: Improper Resolution of Path Equivalence
	CWE-22: Improper Limitation of a Pathname to a Restricted Directory ('Path Traversal')
	CWE-551: Incorrect Behavior Order: Authorization Before Parsing and Canonicalization

Mitigations	
	• Convert all paths to their canonical values
	• Ensure you use the canonical paths before making access control decisions
	• Ensure the files are in the expected folder structure (sub-folders may or may not be acceptable)
	• Use standard libraries to normalize paths to their canonical form

7. of Tampering II

An attacker can bypass permissions because you don't make names canonical before checking access permissions:

Threat	
	You're using email addresses as usernames but are not considering the domain. So, when you have users with the same name working for different organizations, there is a name clash, and they gain access to the other users' accounts.
CAPEC	N/A
ASVS	N/A
CWE	CWE-289: Authentication Bypass by Alternate Name

CWE-290: Authentication Bypass by Spoofing

CWE-551: Incorrect Behavior Order: Authorization Before Parsing and Canonicalization |

Mitigations	
	• Ensure you use the canonical or fully qualified name to avoid clashes

8. of Tampering

An attacker can manipulate data because there's no integrity protection for data on the network:

Threat	
	You're sending payloads to your API in clear text and aren't signing them, so an attacker can intercept the request and change the content without being traced.
CAPEC	CAPEC-94: **Adversary in the Middle (AiTM)** CAPEC-384: Application API Message Manipulation via **Man-in-the-Middle (MiTM)**
ASVS	1.9.1: Ensure you're using TLS everywhere 13.2.6: Ensure the integrity of headers and payloads
CWE	CWE-311: Missing Encryption of Sensitive Data CWE-353: Missing Support for Integrity Check CWE-347: Improper Verification of Cryptographic Signature CWE-471: **Modification of Assumed-Immutable Data (MAID)** CWE-924: Improper Enforcement of Message Integrity During Transmission in a Communication Channel

Mitigations	
	• Sign payload data • Encrypt network traffic with TLS

9. of Tampering I

An attacker can provide or control state information:

Threat	
	You aren't signing or securing your cookies correctly, so an attacker can modify a cookie.
CAPEC	CAPEC-31: Accessing/Intercepting/Modifying HTTP Cookies
ASVS	3.4 and 3.: Ensure cookies are secured properly and only accessible from the source host
CWE	CWE-565: Reliance on Cookies without Validation and Integrity Checking

Mitigations
• Cookies should be signed so that attackers cannot modify the state stored within them
• Cookies should only be communicated over TLS
• Cookies should be domain-restricted so that another domain cannot use/access them
• Cookies should contain the `SameSite`, `Secure`, and `HttpOnly` attributes

9. of Tampering II

An attacker can provide or control state information:

Threat	
	You're accepting session IDs in GET and POST requests and not changing the user's session ID after they have logged in, so an attacker is targeting your users by sending them a URL in a phishing email that already contains the session ID. Once your users are authenticated, the attacker is then able to hijack the users' sessions and gain access to your systems.
CAPEC	CAPEC-61: Session Fixation CAPEC-593: Session Hijacking
ASVS	3.2.1: Ensure that a new session is created on login
CWE	CWE-384: Session Fixation

Mitigations	
	• Convert all paths into their canonical values • Ensure you use the canonical paths before making access control decisions • Ensure the files are in the expected folder structure (sub-folders may or may not be acceptable).

10. of Tampering I

An attacker can alter information in a data store because it has weak ACLs or includes a group that is equivalent to everyone (*all Live ID holders*).

The alternative text is that an attacker can alter information in a data store because it has weak/open permissions or includes a group that is equivalent to everyone (*anyone with a Facebook account*):

Threat	
	You have given everyone full permissions on your database schema and now they can create, read, update, and delete data, or maybe they can modify the schema itself.
CAPEC	CAPEC-180: Exploiting Incorrectly Configured Access Control Security Levels CAPEC-1: Accessing Functionality Not Properly Constrained by ACLs
ASVS	4.1.3: Ensure users or services only have the necessary privileges to perform the actions they need to do
CWE	CWE-921: Storage of Sensitive Data in a Mechanism without Access Control

Mitigations	
	• Ensure you have well-defined roles • Restrict which roles can administer the data • Restrict which roles can administer the database schema • Encrypt sensitive data • Implement the principle of least privilege.

10. of Tampering II

An attacker can alter information in a data store because it has weak ACLs or includes a group that is equivalent to everyone (*all Live ID holders*).

The alternative text is that an attacker can alter information in a data store because it has weak/open permissions or includes a group that is equivalent to everyone (*anyone with a Facebook account*):

Threat	
	The policy associated with your cloud object storage allows write access to any IAM role in the organization.
CAPEC	CAPEC-180: Exploiting Incorrectly Configured Access Control Security Levels CAPEC-1: Accessing Functionality Not Properly Constrained by ACLs
ASVS	4.1.3: Ensure users or services only have the necessary privileges to perform the actions they need to do
CWE	CWE-921: Storage of Sensitive Data in a Mechanism without Access Control

Mitigations	
	• Encrypt sensitive data • Implement the principle of least privilege • Implement file/volume-level encryption • Always use secure defaults and relax controls as needed

Jack of Tampering

An attacker can write to some resource because permissions are granted to the world or there are no ACLs:

Threat	
	You have a file stored on a Unix-based operating system that has the permissions set to 777 or -rwxrwxrwx. This means that you, anyone in the same group as you, or anyone with access to the machine has read, write, and execute access.
CAPEC	CAPEC-576: Group Permission Footprinting CAPEC-180: Exploiting Incorrectly Configured Access Control Security Levels
ASVS	4.1.3: Ensure users or services only have the necessary privileges to perform the actions they need to do
CWE	CWE-552: Files or Directories Accessible to External Parties CWE-285: Improper Authorization CWE-668: Exposure of Resource to Wrong Sphere CWE-732: Incorrect Permission Assignment for Critical Resource

Mitigations	
	• Implement the principle of least privilege when setting permissions on files • Always use secure defaults and relax controls as needed

Queen of Tampering

An attacker can change parameters over a trust boundary and after validation (for example, important parameters in a hidden field in HTML, or passing a pointer to critical memory):

Threat	
	You're not performing any validation or sanitization on the backend before processing the data. This leaves you open to an attacker using a proxy such as ZAP to capture the request and modify the parameters after any validation and/or sanitization has been performed in the browser, thus bypassing your protective measures.
CAPEC	CAPEC-74: Manipulating State CAPEC-157: Sniffing Attacks CAPEC-77: Manipulating User-Controlled Variables CAPEC-384: Application API Message Manipulation via Man-in-the-Middle CAPEC-248: Command Injection CAPEC-66: SQL Injection
ASVS	1.5.1: Ensure that input and output requirements are clearly defined 1.5.3: Ensure that validation is performed where the data can't subsequently be tampered with
CWE	CWE-137: Data Neutralization Issues CWE-1215: Data Validation Issues

Mitigations	
	• Ensure you validate the data on the backend • Ensure you sanitize the data on the backend

King of Tampering I

An attacker can load code inside your process via an extension point:

Threat	
	Your application allows extensions to be uploaded at runtime – for example, a binary or script or even some XML that instructs the machine at runtime. That extension causes the application to behave in an undesired way or even lead to a data disclosure.
CAPEC	CAPEC-698: Install Malicious Extension.
ASVS	10.3.2: Ensure the application doesn't load code from untrusted sources.
CWE	CWE-434: Unrestricted Upload of File with Dangerous Type CWE-494: Download of Code Without Integrity Check

Mitigations	
✓	• Require authentication for loading plugins • Plugins must be signed

King of Tampering II

An attacker can load code inside your process via an extension point:

Threat	
	You're using object deserialization in Java without verifying the source or ensuring that the content is of the expected type before deserializing it. As a consequence, your system can inadvertently be runtime-loading executable code.
CAPEC	CAPEC-548: Contaminate Resource CAPEC-242: Code Injection
ASVS	5.5.3 – Ensure you are only allowing deserialization of objects you are allow listing
CWE	CWE-502: Deserialization of Untrusted Data

Mitigations	
	• Use allow listing to ensure the data being deserialized is of an expected type • Avoid deserialization of untrusted sources

Ace of Tampering I

You've invented a new tampering attack:

Threat	
	You're using deny listing to restrict data that will be processed by your application or avoid injection attacks and an attacker uses an alternate encoding to bypass your validation because they haven't thought of this edge case.
CAPEC	CAPEC-267: Leverage Alternate Encoding
ASVS	5.1.4: Ensure a schema is used when validating XML and JSON data.
CWE	CWE-791: Incomplete Filtering of Special Elements

Mitigations	
	• The preferred solution should always be to use allow listing where possible
	• Verify that the character set is the one you expect
	• Restrict the alphabet

Ace of Tampering II

You've invented a new tampering attack:

Threat	
	An attacker adds additional fields to a payload being passed to an API call. You process all key-value pairs and your default behavior doesn't validate the input.
CAPEC	CAPEC-36: Using Unpublished Interfaces or Functionality CAPEC-138: Reflection Injection
ASVS	5.1.1: Ensure parameter pollution safeguards are in place 5.1.2: Ensure only allowed fields are retrieved from the payload 5.1.3: Validate all untrusted data and use allow lists where possible
CWE	CAPEC-235: Improper Handling of Extra Parameters CAPEC-915: Improperly Controlled Modification of Dynamically-Determined Object Attributes

Mitigations	
	• Allowlist or only read the expected fields • Use a schema document to validate the input structure before processing the data • The default behavior should be secure

E of Tampering I

Data in the database can be *fixed* by the admins, and nobody will ever know:

Threat	
	Administrators have change access to the database for CRUD operations but there is no audit trail or approval process, so they can make changes without detection.
CAPEC	N/A
ASVS	7.1.3: Ensure security events are being logged 7.1.4: Ensure log entries contain all the necessary information for an investigation
CWE	CWE-778: Insufficient Logging

Mitigations	
	• There should be an audit trail of changes being made to the data so that it is possible to trace who changed what • Changes to data being performed by an administrator should require approval from a second party so that nefarious actions require collusion • Ensure that you conduct proper security background checks on any personnel with access to sensitive data

E of Tampering II

Data in the database can be *fixed* by the admins, and nobody will ever know:

Threat	
	Your administrators can modify their permissions to grant themselves read/write access to data without secondary approval.
CAPEC	N/A
ASVS	7.1.3: Ensure security events are being logged
	7.1.4: Ensure log entries contain all the necessary information for an investigation
	7.2.1: Ensure both authentication successes and failures are logged
	7.2.2: Ensure both authorization successes and failures are logged
CWE	CWE-778: Insufficient Logging

Mitigations	
	• Require separation of duties for administrators to gain access to data
	• Implement the principle of least privilege

Summary

Now that we've covered the Tampering suit from the Elevation of Privilege card deck, you should be familiar with the different types of tampering threats and what the design flaws might look like that leave you open to these threats in a software design.

Tampering can often have serious consequences and can allow an attacker to change settings or state information, giving them additional access rights or permissions so that they can perform some other action. As you work through this book, on occasion, you will see that there are blurred lines as to which category a threat might belong to. This is because threats are often linked, and one threat is made possible by the existence of another. How you classify these threats isn't as important as detecting them and understanding their implications and how to protect against them.

Being able to recognize and mitigate tampering threats can help protect you from other threats that might have been possible as a consequence.

In the next chapter, we're going to look at the repudiation threats described on the cards in the Repudiation suit in the Elevation of Privilege card deck.

4
Repudiation

Repudiation is plausible deniability, or rather the inability to prove that someone did something. When you think about repudiation, you should think about threats that affect your ability to hold people accountable. Three things are required for an action to be non-repudiable: the what, the who, and the when, and this information should be immutable.

Figure 4.1: Destroying the logs/evidence

In this chapter, we will cover the threats described in the Repudiation suit in the Elevation of Privilege card deck, including an additional four cards from the T.R.I.M. extension to the game. We'll go through some examples of repudiation threats; I'll give you references with each example where you can get more information and I will also suggest what mitigations and controls you can put in place to protect against the threat or at least reduce the risk.

By the end of the chapter, you'll have a better understanding and awareness of what the different repudiation threats are and be able to recognize them in your system architecture and designs. Before we do, let's look at why repudiation is important and how it relates to security.

The importance of repudiation and its role in security

Your logging subsystem may be crucial in giving you early indicators of compromise. By using anomaly detection, you can trigger alerts when something unusual is taking place that needs investigation. By feeding all of your logs into a central log server or a **Security Information and Event Management** (**SIEM**) system, you can configure rules that will trigger these alerts when certain conditions hold true. In the event of a security incident, log information can be of the utmost importance because it can help you determine what happened, how it happened, when it happened, and what was affected. The benefits of this are as follows:

- It will allow you to perform a root-cause analysis

- It will help you remediate the cause to reduce the risk that it will happen again in the future

- It may be a requirement that you inform anyone affected by the incident

- Logs may be needed as evidence if this was malicious activity

Due to the forensic nature of logs, they should be immutable; you should only be able to append to the logs but never modify or delete from the logs so that bad actors can't cover their tracks. You must be able to show what actions were carried out, who performed those actions, when they performed those actions, and on what they performed those actions. You may want to keep an audit trail, or you may store the information in your log files.

If someone's data is leaked, we will need to show who accessed that data, for what reason, and when. Audit logs will then help when investigating how the data was leaked. It may be that one or more of your accounts have been compromised.

Logs themselves can also be the target of an attack if they contain sensitive information. So, how you store the information and the level of security used should be commensurate with the security classification of the data you are logging. If you are logging **personally identifiable information** (**PII**) that should only be visible to someone in human resources, then access to the log file should also be restricted to members of human resources.

Now that we have covered the importance of repudiation and what its role is in secure architecture, let's start looking into the cards.

2. of Repudiation

An attacker can pass data through the log to attack a log reader, and there's no documentation of what sort of validations are done.

Threat	
	You are not sanitizing user-supplied data before logging it. So, an attacker can inject JavaScript into the logs so that when the logs are read from a browser-based log reader, the code that was stored in the log is executed.
CAPEC	CAPEC-592 - Stored XSS
ASVS	5.3 - Ensure output is being encoded and protection is in place against injection attacks.
	7.3.1 - Verify that logs are being encoded to avoid **Carriage Return Line Feed** (**CRLF**) injection, stored **Cross-Site Scripting** (**XSS**), and other forms of log injection.
CWE	CWE-117 - Improper Output Neutralization for Logs
	CWE-94 - Improper Control of Generation of Code (*Code Injection*)

Mitigations	
	• Ensure that logs are sanitized before being stored
	• Ensure that logs being read are sanitized before being output to the log reader
	• Document what is being validated or sanitized so that clients can implement additional protection for anything missing

3. of Repudiation

A low-privilege attacker can read interesting security information in the logs.

Threat	
	You don't have any access control in place to restrict who can access the logs; permissions to access the logs should be the same as those required to view any of the information being logged.
CAPEC	CAPEC-180 - Exploiting Incorrectly Configured Access Control Security Levels
ASVS	7.1.1 - Ensure that secrets and payment card details including CVV numbers are not being logged. 7.1.2 - Ensure PII and other sensitive data being logged complies with regulations. 7.3.3 - Ensure logs are protected by strict access controls and that the entries are immutable.
CWE	CWE-921 - Storage of Sensitive Data in a Mechanism without Access Control. CWE-1220 - Insufficient Granularity of Access Control.

Mitigations	
	• Implement the principle of least privilege: a user should only have the privileges necessary to perform their intended function. If they require more privileges for a specific task, they should be granted for the length of time necessary to perform that task. Define which roles should have access to logs. • Classify your logs and grant access to roles based on their classification. • Don't put sensitive information into the logs. • Encrypt sensitive data in the logs.

4. of Repudiation

An attacker can alter digital signatures because the digital signature system you're implementing is weak or uses MACs where it should use a signature.

Threat	
	Your system uses shared keys to generate **message authentication codes** (**MACs**). These shared keys mean that both the sender and the receiver have the same key and, therefore, you cannot rely on this as a means to identify the source of the message. This poses a number of issues. As both you (the sender) and the receiver need the key, you will need a secure channel on which to share the key. You will also need to have a key for every recipient of the message and create a separate signature for each of them.
CAPEC	CAPEC-151 - Identity Spoofing CAPEC-195 - Principal Spoof CAPEC-194 - Fake the Source of Data
ASVS	1.8.2 - Ensure you have requirements for each classification of your data for each category of the CIA triad (confidentiality, integrity, and availability).
CWE	CWE-322 - Key Exchange without Entity Authentication

Mitigations	
	• Digitally sign your messages using PKI • Use a strong hashing algorithm • To make your messages valid in court, have your signature signed by a third-party timestamping service

5. of Repudiation

An attacker can alter log messages on a network because they lack strong integrity controls.

Threat	
	You are not verifying the source of log information, your network traffic isn't encrypted, and you aren't signing log messages, so an attacker can intercept and modify them, and you will never know.
CAPEC	CAPEC-94 - Adversary in the Middle (AiTM) CAPEC-216 - Communication Channel Manipulation CAPEC-217 - Exploiting Incorrectly Configured SSL/TLS
ASVS	1.9.1 - Ensure you're using TLS everywhere. 9.1.1 - Ensure the TLS version can't be downgraded. 9.1.3 - Ensure up-to-date versions of the TLS protocol are used. 9.2.1 - Ensure that TLS handshakes are performing certificate verification. 9.2.3 - Ensure you are authenticating and encrypting any sensitive traffic.
CWE	CWE-924 - Improper Enforcement of Message Integrity During Transmission in a Communication Channel

Mitigations	
✅	• Sign log messages • Use MTLS to verify the source • Authenticate the sender

6. of Repudiation I

An attacker can create a log entry without a timestamp (or no log entry is timestamped).

Threat	
	You aren't sanitizing CRLFs before writing data from untrusted sources to the logs. Using CRLF injection, an attacker can insert additional log lines by adding a new line before inserting some text they have generated to fake one or more additional log lines.
CAPEC	CAPEC-93 - Log Injection-Tampering-Forging
ASVS	5.3 - Ensure output is encoded and you're protecting against injection attacks. 7.3.4 - Ensure clocks are synched with an atomic clock and are in the correct time zone.
CWE	CWE-93 - Improper Neutralization of CRLF Sequences ('CRLF Injection')

Mitigations
• If using Log4j, use the encode CRLF format pattern to sanitize log messages • Use an encoder library such as the OWASP Encoder

6. of Repudiation II

An attacker can create a log entry without a timestamp (or no log entry is timestamped).

Threat	
	Your log appender pattern doesn't include a timestamp. Timestamps are important because they allow you to correlate logs between different parts of your system, but they also allow you to see whether log entries are missing because there would be a break in the continuity of the entries.
CAPEC	CAPEC-268 - Audit Log Manipulation
ASVS	7.1.4 - Ensure log entries are sufficiently complete for forensics to reconstruct the chain of events.
CWE	CWE-223 - Omission of Security-Relevant Information

Mitigations	
	• Add a timestamp to your log appender pattern

7. of Repudiation

An attacker can make the logs wrap around and lose data.

Threat	
	Your log appender may be set to overwrite when the log gets too large, and the attacker is forcing your application to generate too much log data deliberately.
CAPEC	CAPEC-268 - Audit Log Manipulation CAPEC-81 - Web Server Logs Tampering
ASVS	N/A
CWE	CWE-222 - Truncation of Security-Relevant Information

Mitigations	
	• Ensure that your logging isn't excessive, creating so much noise that it becomes impossible to see important security information in the log • Configure your logging subsystem (the library you use for logging, for example, Log4j) correctly • Ensure you have sufficient disk space • Log to a central system or data lake

8. of Repudiation

An attacker can make the logs lose or confuse security information.

Threat	
	You are not sanitizing untrusted data going into the logs. An attacker might perform a log injection of newline characters to be able to complete a log line, thus making it look as though they performed some innocuous tasks while writing fake log lines afterward that contain the actual task performed. This action makes it appear that someone else is responsible.
CAPEC	CAPEC-93 - Log Injection-Tampering-Forging CAPEC-268 - Audit Log Manipulation CAPEC-81 - Web Server Logs Tampering
ASVS	5.3 - Ensure output is encoded and you're protecting against injection attacks. 7.3.4 - Ensure clocks are synched with an atomic clock and are in the correct time zone.
CWE	CWE-93 - Improper Neutralization of CRLF Sequences ('CRLF Injection')

Mitigations	
	• Sanitize the messages you are about to log if they contain untrusted data • Depending on your logging subsystem, encoding of CRLF in the log pattern

9. of Repudiation

An attacker can use a shared key or authenticate as different principals, confusing the information in the logs.

Threat	
	You have multiple services using the same service account so an attacker can steal the credentials or token of the service account and then use it to access different parts of the system.
CAPEC	CAPEC-151 - Identity Spoofing CAPEC-195 - Principal Spoof CAPEC-194 - Fake the Source of Data
ASVS	2.2.5 - Ensure identity and access management components authenticate with each other via Mutual TLS. 2.10.1 - Ensure service-to-service auth doesn't use static tokens. 2.10.4 - Ensure secrets are handled and stored securely, and never hardcoded in your applications.
CWE	CWE-322 - Key Exchange without Entity Authentication

Mitigations	
	• Ensure that you are adequately protecting your stored credentials in a **privileged access management (PAM)** system • Ensure that you are rotating those controls regularly • Use IP restrictions to ensure that those credentials can only be used from expected servers • Implement revocation lists that are kept updated, so that if a token has not yet expired, it can still be revoked and a session invalidated, for example

10. of Repudiation

An attacker can get arbitrary data into logs from unauthenticated (or weak authenticated) outsiders without validation.

Threat	
	You've centralized your logging, but the loggers aren't authenticated with the central system, so anyone could fake the logs for some part of the system.
CAPEC	CAPEC-194 - Fake the Source of Data CAPEC-115 - Authentication Bypass CAPEC-22 - Exploiting Trust in Client
ASVS	1.2.2 - Ensure access to the Data layer is authenticated and the principle of least privilege is being used. 7.3.3 - Ensure logs are protected by strict access controls and that the entries are immutable.
CWE	CWE-921 - Storage of Sensitive Data in a Mechanism without Access Control CWE-1220 - Insufficient Granularity of Access Control

Mitigations
• Use MTLS to authenticate the source • Sign log messages

Jack of Repudiation

An attacker can edit logs and there's no way to tell (perhaps because there's no heartbeat option for the logging system).

Threat	
	You are not signing or adding a sequence number to log entries, so they can either be deleted or changed and you would never know.
CAPEC	CAPEC-81 - Web Server Logs Tampering CAPEC-268 - Audit Log Manipulation
ASVS	N/A
CWE	CWE-353 - Missing Support for Integrity Check

Mitigations	
	• Signing log messages would protect them from being changed • Giving them a sequence number that is also part of the signature ensures that they cannot be deleted either • Add a heartbeat so that any time spent manually changing logs would cause there to be missing heartbeat entries

Queen of Repudiation I

An attacker can say, "*I didn't do that*," and you'd have no way to prove them wrong.

Threat	
	You are not putting usernames or some other identifier into your audit trail entries, so you can't trace an action that was performed back to the individual who performed it.
CAPEC	N/A
ASVS	7.1.3 - Ensure security events are being logged.
CWE	CWE-778 - Insufficient Logging CWE-223 - Omission of Security-Relevant Information

Mitigations	
	• Include identifiers for users in log messages • Ensure the user being logged is the user who performed the action and not the technical user used to launch the application

Queen of Repudiation II

An attacker can say, "*I didn't do that*," and you'd have no way to prove them wrong.

Threat	
	You are not synchronizing system time between environments, so it is impossible to correlate data to determine who did what across the entire system. If your software architecture includes multiple services talking to one another, you may need to look at log files for multiple different services to trace what happened as data moved through your system. If those services are not synchronized, it becomes difficult to connect the log entries to the same data flow.
CAPEC	N/A
ASVS	7.1.4 - Ensure log entries are sufficiently complete for forensics to reconstruct the chain of events. 7.3.4 - Ensure clocks are synched with an atomic clock and are in the correct time zone.
CWE	N/A

Mitigations	
	• Synchronize the time across all systems using NTP

King of Repudiation I

The system has no logs.

Threat	
	You have logging turned off to save disk space.
CAPEC	N/A
ASVS	7.1.3 - Ensure security events are being logged.
CWE	CWE-778 - Insufficient Logging CWE-779 - Logging of Excessive Data CWE-222 - Truncation of Security-Relevant Information

Mitigations	
	• Turn logging on • Log to a remote server

King of Repudiation II

The system has no logs.

Threat	
	Your application doesn't have write access to the folder where logs should be written.
CAPEC	N/A
ASVS	7.1.3 - Ensure security events are being logged.
CWE	CWE-778 - Insufficient Logging CWE-222 - Truncation of Security-Relevant Information

Mitigations	
	• Ensure your application has write access to the folder where logs should be stored

King of Repudiation III

The system has no logs.

Threat	
	You haven't implemented logging in your code.
CAPEC	N/A
ASVS	7.1.3 - Ensure security events are being logged.
	7.2.1 - Ensure both authentication successes and failures are logged.
	7.2.2 - Ensure both authorization successes and failures are logged.
CWE	CWE-778 - Insufficient Logging

Mitigations	
	• Implement logging of relevant events in your code with different levels of logging for different classifications of log messages
	• Identify relevant security events for logging with your application team
	• Ensure you are not logging so much information that it becomes too difficult to find the relevant information

Ace of Repudiation

You've invented a new repudiation attack.

Threat	
	Your central logging server is a single point of failure, and an attacker can use **Domain Name Server (DNS)** or **Address Resolution Protocol (ARP)** poisoning to render your centralized log server unreachable.
CAPEC	CAPEC-571 - Block Logging to Central Repository

CAPEC-589 - DNS Blocking

CAPEC-598 - DNS Spoofing |
| ASVS | 10.3.3 - Ensure your DNS is kept up to date and protected from subdomain takeovers. |
| CWE | N/A |

Mitigations	
	• Use **Domain Name System Security Extensions (DNSSEC)** on your local network

• Ensure that your applications validate DNS record signatures for local network addresses

• Use TLS and verify the server certificate

• Alert when IP clashes occur on the network

• Retain a buffer of log information so that, in the event of an outage, you have a window in which to recover

• Queue logs until the remote server is available |

E. of Repudiation

We don't log personal data access because we do not process any customer data, only employee data.

Threat	
	Your employees' data may contain PII that is even more sensitive than the data you are storing about your customers. Employees also have citizen rights just like third parties and, as such, their data should be afforded the same protection as customers' data with the same level of sensitivity.
CAPEC	CAPEC-150 - Collect Data from Common Resource Locations CAPEC-155 - Screen Temporary Files for Sensitive Information
ASVS	1.8.1 - Ensure all your data is given a classification. 1.8.2 - Ensure you have requirements for each classification of your data for each category of the CIA triad (confidentiality, integrity, and availability). 7.1.2 - Ensure PII and other sensitive data being logged complies with regulations. 7.3.3 - Ensure logs are protected by strict access controls and that the entries are immutable. 8.3.5 - Ensure you have an audit trail for all sensitive data access.
CWE	CWE-779 - Logging of Excessive Data CWE-532 - Insertion of Sensitive Information into Log File

Mitigations	
	• Always log access to any personal data • Access to personal data should be restricted based on the role of the person seeking access and the personal data being accessed • Review the organizational data handling policies and procedures and, where necessary, make adjustments to the framework

F. of Repudiation

We log changes and deletions of personal data, but viewing is not logged.

Threat	
	You may not want your colleagues to know where you live, your phone number, or your salary. However, this information is written to logs that they have access to.
CAPEC	N/A
ASVS	1.8.1 - Ensure all your data is given a classification. 1.8.2 - Ensure you have requirements for each classification of your data for each category of the CIA triad (confidentiality, integrity, and availability). 7.1.2 - Ensure PII and other sensitive data being logged comply with regulations. 7.3.3 - Ensure logs are protected by strict access controls and that the entries are immutable. 8.3.5 - Ensure you have an audit trail for all sensitive data access.
CWE	CWE-779 - Logging of Excessive Data CWE-532 - Insertion of Sensitive Information into Log File CWE-215 - Insertion of Sensitive Information into Debugging Code

Mitigations	
	• Access to this information should be on a need-to-know basis • Always log access to any personal data • Log access should be classified at the same level as the most sensitive piece of information being logged • Where possible, don't log PII

G. of Repudiation

We log personal data access, but there is no ongoing monitoring or alerting.

Threat	
	One of your staff has been reading or exporting data about other employees. You have just discovered this, but it has been going on for 6 months.
CAPEC	N/A
ASVS	1.7.2 - Ensure logs are sent securely to a remote server or to a Security Information and Event Management (SIEM) system.
CWE	CWE-215 - Insertion of Sensitive Information into Debugging Code

Mitigations	
	• Using a SIEM system, you could configure it to detect when people are accessing sensitive data without good reason • Configure the SIEM system to send notifications

H. of Repudiation I

Our audit log contains personal data, and we do not record who looks at our audit logs.

Threat	
	You are not reviewing what should be logged and are logging personal data when you should not.
CAPEC	N/A
ASVS	7.1.1 - Ensure that secrets and payment card details including CVV numbers are not being logged. 7.1.2 - Ensure PII and other sensitive data being logged complies with regulations. 7.1.3 - Ensure security events are being logged. 8.3.5 - Ensure you have an audit trail for all sensitive data access.
CWE	CWE-215 - Insertion of Sensitive Information into Debugging Code

Mitigations	
	• Perform code reviews to ensure that personal data is not being logged unnecessarily • Implement least privilege access to logs • Record who accesses production logs when and why

H. of Repudiation II

Our audit log contains personal data, and we do not record who looks at our audit logs.

Threat	
	You are not verifying that your configuration settings are secure and have debug enabled in production. The production logs are accessible by everyone in the development team without any approval process.
CAPEC	CAPEC-180 - Exploiting Incorrectly Configured Access Control Security Levels CAPEC-176 - Configuration/Environment Manipulation CAPEC-75 - Manipulating Writeable Configuration Files
ASVS	14.1.3 - Ensure configuration is hardened in line with vendor guidelines. 14.1.5 - Ensure you have integrity controls on configuration and alerting on change events.
CWE	CWE-15 - External Control of System or Configuration Setting CWE-489 - Active Debug Code CWE-215 - Insertion of Sensitive Information into Debugging Code

Mitigations	
	• Perform configuration and change reviews to ensure you have the correct and secure configuration settings for each environment • Implement least privilege access to logs • Record who accesses production logs when and why

Summary

You've now covered the threat types described on the cards from the Repudiation suit in the Elevation of Privilege card deck with the addition of four cards from the T.R.I.M. extension for the game. These threats detailed flaws relating to insufficient logging, excessive logging, logging the wrong content, integrity of the logs, file permissions, and network connectivity.

You should now understand how the confidentiality, integrity, and availability of logs can be affected by a variety of different threats. Having a greater awareness of the threats in this category and having discussed how to mitigate them, you are now armed with the tools you need to ensure that, if something does go wrong, you have what you need to audit the events that led up to the issue.

In the next chapter, we will cover the fourth S.T.R.I.D.E. category of threats, Information Disclosure.

5
Information Disclosure

Confidentiality is often fundamental and information disclosure is when that confidentiality is compromised in some way. Your data is probably your most precious asset; that data might be **personally identifiable information (PII)** such as the names and addresses of your customers, it might be trade secrets such as a recipe or it could be the company's finances. Whatever that data is, it needs to be protected adequately and, in this chapter, we look at some of the common threats in this category and how you can mitigate them.

Figure 5.1: Information being leaked

We will start this chapter by briefly looking at some key concepts around password management, key management, and cryptography. We will then cover the threats described on the cards from the Information Disclosure suit in the Elevation of Privilege card deck, including an additional two cards from the T.R.I.M. extension to the game. We'll go through some examples of information disclosure threats, and I'll give you references with each example where you can get more information. References are not necessarily available in all the directories; for example, MITRE CWE and OWASP ASVS do not include social engineering entries. I will also suggest what mitigations and controls you can put in place to protect against the threat or at least reduce the risk.

By the end of the chapter, you'll have a better understanding and awareness of what the different information disclosure threats are and be able to recognize them in your system architecture and designs.

Understanding password management, key management, and cryptography

Before we jump straight in, authentication, authorization, and cryptography are fundamental to protecting confidentiality, so just a few words on password management, key management, and cryptography. There are many books dedicated to these topics that go into much more detail than I ever could here, but I'll cover some of the topics quickly to give you some context as we go through the chapter.

Authentication is a mechanism for verifying that a person is who they say they are, and authorization is a mechanism for determining whether they have sufficient privileges for the object they are trying to access.

Password management

Using the same password for different accounts is an enormous risk because, if a site where it was used gets breached and it wasn't stored securely, then it is probable an attacker will try and reuse that username and password combination on other sites. Short passwords of eight letters or less, and even complex ones, can be broken in less than 8 hours. Attackers will first try the most commonly used passwords (which are published regularly on the internet). They will also use dictionaries containing pet names, first names, football teams, and other words commonly used in passwords to try first, maybe in combination with dates.

When passwords are stored for authentication, they are often stored in a hashed form. This is a value that is calculated using a mathematical function that should be irreversible. The function will generate a unique output value for every unique input value but will always generate the same output given the same input value. By adding a small number of random characters before generating the hash, it will generate a completely different value. This is called salting.

Salting is used because, if an attacker knows the algorithm and no random characters were added, the attacker can precalculate a list of hashes, known as rainbow tables, with which to compare against the hash of your password, allowing them to quickly find your password. At the same time, if a site always uses the same salt, then an attacker (after discovering one password) can quickly discover the salt used and precompile their list of hashes to compare for all the other credentials.

Key management

Keys should only be trusted if you are certain the organization that is using them is who they claim to be. This is why we have something called a **Certificate Authority** (**CA**). They are responsible for verifying the identity of individuals or organizations and their legitimacy before issuing a key to them. When they issue the key to them, this key will be digitally signed with a key belonging to the CA and the key will also have an expiry, usually 1 year.

Keys, like passwords, should be rotated regularly and immediately if you suspect they have been compromised, and like passwords, keys can be stolen. For this reason, CAs maintain lists of keys that have been revoked. Browsers usually come with a set of pre-installed public keys with which to verify the CA signatures of keys used by websites to allow you to verify their authenticity.

A certificate is also usually associated with the fully qualified domain name of a website and contains additional information about the organization. If the certificate is used on another host, the browser will notify you that the certificate is invalid.

Cryptography

There are two main types of cryptography: symmetric and asymmetric cryptography.

Symmetric cryptography requires the same key to both encrypt and decrypt information. It is faster than asymmetric cryptography but suffers from a fundamental problem: both the sender and the receiver have a shared copy of the key. The problem here is how to exchange the key in the first place without someone intercepting it and that it cannot be used to guarantee the authenticity of the source of the information.

Asymmetric cryptography is slower than symmetric cryptography and uses a key pair: one key to encrypt and another key to decrypt. So, if I want someone to send me some information that only I can read, I can share my public key with them over any channel because only I can decrypt any information encrypted with it. After all, only I have the private key.

If I want to prove that I sent something, I can hash the information and then encrypt the hash with my private key creating a digital signature. Anyone with my public key can then hash the same information, decrypt the signature with my public key to retrieve the hash, and compare them. If they are the same, this proves that the information hasn't been tampered with and that I was the source.

Now that we have covered some relevant details about password management, key management, and cryptography, let's start looking into the cards.

2. of Information Disclosure

An attacker can brute-force file encryption because there's no defense in place (example defense: password stretching).

Threat	
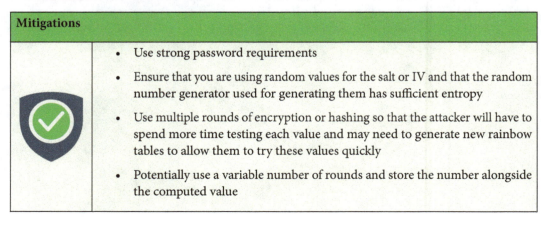	The effort required to break your encryption/hashing algorithm is relatively small because your minimum requirements for passphrase/password strength aren't perhaps as strong as they should be, or you are using the same initialization vector/salt for them all. Therefore, the attacker can try passwords very quickly, perhaps using rainbow tables (a pre-hashed or encrypted list of commonly used passwords) because there is nothing to slow them down.
CAPEC	CAPEC-112 - Brute Force CAPEC-20 - Encryption Brute Forcing CAPEC-16 - Dictionary-based Password Attack CAPEC-55 - Rainbow Table Password Cracking
ASVS	2.1 - Ensure the quality of Password Policy implementation 2.2 - Ensure the quality of authentication factors
CWE	CWE-261 - Weak Encoding for Password CWE-328 - Use of Weak Hash CWE-331 - Insufficient Entropy CWE-335 - Incorrect Usage of Seeds in **Pseudo-Random Number Generator** (PRNG) CWE-1204 - Generation of Weak **Initialization Vector** (IV)

Mitigations	
	• Use strong password requirements • Ensure that you are using random values for the salt or IV and that the random number generator used for generating them has sufficient entropy • Use multiple rounds of encryption or hashing so that the attacker will have to spend more time testing each value and may need to generate new rainbow tables to allow them to try these values quickly • Potentially use a variable number of rounds and store the number alongside the computed value

3. of Information Disclosure I

An attacker can see error messages with security-sensitive content.

Threat	
	The system will tell the attacker whether the username is incorrect, or the password is incorrect. So, the attacker will know they have found a valid user.
CAPEC	CAPEC-112 - Brute Force CAPEC-575 - Account Fingerprinting CAPEC-70 - Try Common or Default Usernames and Passwords CAPEC-565 - Password Spraying
ASVS	7.4.1 - Ensure error messages don't leak security-sensitive information to the user.
CWE	CWE-209 - Generation of Error Messages Containing Sensitive Information CWE-210 - Self-Generated Error Message Containing Sensitive Information

Mitigations	
	• Always return the same message, for example, "username or password incorrect" without distinguishing which one, even when a user does not exist. • Return an error ID with the message, which allows support staff to request the ID from the customer and cross-reference the logs to determine the cause. The ID should be a random value that changes for every error.

3. of Information Disclosure II

An attacker can see error messages with security-sensitive content.

Threat	
	An attacker tries to retrieve a report before verifying that the user has sufficient access rights. It fails and the reason you give isn't related to their permissions but perhaps tells them information relating to the report, such as its security classification. You are, therefore, confirming its existence to the attacker.
CAPEC	CAPEC-497 - File Discovery
	CAPEC-694 - System Location Discovery
	CAPEC-577 - Owner Footprinting
	CAPEC-576 - Group Permission Footprinting
ASVS	7.4.1 - Ensure error messages don't leak security-sensitive information to the user.
CWE	CWE-209 - Generation of Error Messages Containing Sensitive Information
	CWE-210 - Self-Generated Error Message Containing Sensitive Information

Mitigations	
	• Perform your authentication checks upfront
	• Perform your authorization checks upfront
	• Implement secure by default, fail fast, and fail secure

4. of Information Disclosure

An attacker can read content because messages (say, an email or HTTP cookie) aren't encrypted even if the channel is encrypted.

Threat	
	An administrator has access to the mailboxes on the server and can, therefore, read the content of your emails even though, between the sender and them arriving at your mailbox, they were transmitted over a secure channel.
CAPEC	CAPEC-180 - Exploiting Incorrectly Configured Access Control Security Levels
ASVS	8.3.5 - Ensure you have an audit trail for all sensitive data access. 8.3.7 - Use appropriate levels of encryption for the classification of the data.
CWE	CWE-312 - Cleartext Storage of Sensitive Information

Mitigations	
	• Using S/MIME in this case would encrypt the content of the mail, and even if someone had access to your mailbox on the server, they still wouldn't be able to read the content

5. of Information Disclosure

An attacker may be able to read a document or data because it's encrypted with a non-standard algorithm.

Threat	
	You're using an encryption you wrote yourself to protect your data instead of a known strong encryption algorithm that has been industry-tested.
CAPEC	CAPEC-192 - Protocol Analysis CAPEC-97 - Cryptanalysis
ASVS	6.2.2 - Ensure standard recognized and tested cryptographic algorithms are being used.
CWE	CWE-327 - Use of a Broken or Risky Cryptographic Algorithm

Mitigations	
	• Encryption is very hard to implement, and most people get it wrong. Ensure the use of industry-recognized standard libraries. • Use the strongest, most suitable algorithm you can at the time. • Make your solution cryptographically agile. You can migrate to a new algorithm without too much difficulty if the algorithm you are using gets broken.

6. of Information Disclosure

An attacker can read data because it's hidden or occluded (for undo or change tracking) and the user might forget that it's there.

Threat	
	The document format you're using stores information in the file about changes you've made. By examining the file, things that you may have removed because you didn't want the recipient to see them can be recovered.
CAPEC	CAPEC-116 - Excavation CAPEC-150 - Collect Data from Common Resource Locations CAPEC-212 - Functionality Misuse
ASVS	8.1.2 - Ensure proper housekeeping is being performed and that temporary data is cleaned up properly. 8.2.2 - Ensure that data in the frontend and middleware is also cleaned up properly. 8.3.6 - Ensure that variable data stored in memory is dereferenced and destroyed when you've finished processing it.
CWE	CWE-459 - Incomplete Cleanup CWE-922 - Insecure Storage of Sensitive Information CWE-921 - Storage of Sensitive Data in a Mechanism without Access Control CWE-312 - Cleartext Storage of Sensitive Information

Mitigations	
	• Sanitize the file before sharing • Convert to a file type that doesn't support change tracking before sharing

7. of Information Disclosure

An attacker can act as a "man in the middle" because you don't authenticate endpoints of a network connection.

Threat	
	Although you are encrypting the channel, you aren't verifying the hostname against the certificate, you aren't verifying whether the certificate is self-signed or signed by a known CA, you aren't verifying whether the certificate has expired, and you aren't verifying whether the certificate has been revoked.
CAPEC	CAPEC-616 - Establish Rogue Location CAPEC-543 - Counterfeit Websites CAPEC-459 - Creating a Rogue Certification Authority Certificate CAPEC-479 - Malicious Root Certificate CAPEC-384 - Application API Message Manipulation via Man-in-the-Middle
ASVS	9.2.1 - Ensure you're verifying TLS certificates and internal certificate authority certificates or self-signed certificates are also in your trust store. 9.2.4 - Ensure you're also checking whether certificates have been revoked.
CWE	CWE-295 - Improper Certificate Validation CWE-347 - Improper Verification of Cryptographic Signature CWE-346 - Origin Validation Error CWE-941 - Incorrectly Specified Destination in a Communication Channel

Mitigations	
	• Verify that the certificate has been signed by a reputable CA • Verify that the certificate is in your trust store if it is self-signed • Verify that the hostname matches the certificate • Verify that the certificate has not expired • Verify that the certificate has not been revoked • Perform the preceding on both the client and server when using **Mutual Transport Layer Security (mTLS)**

8. of Information Disclosure I

An attacker can address information through a search indexer, logger, or other such mechanism.

Threat	
	You've exposed some files on your web server to bots unwittingly and the contents of those files are now available in the search indexes of different search engines and their caches. They may also be available on the "Wayback Machine."
CAPEC	CAPEC-127 - Directory Indexing CAPEC-143 - Detect Unpublicized Web Pages CAPEC-144 - Detect Unpublicized Web Services
ASVS	4.1.3 - Ensure users or services only have the necessary privileges to perform the actions they need to do.
CWE	CWE-1230 - Exposure of Sensitive Information through Metadata CWE-524 - Use of Cache Containing Sensitive Information

Mitigations	
	• Ask the search engine to purge/remove the data from their cache • Depending on the content that was exposed, rotate any secrets • Notify the data protection authority, within the time frame defined with regulators from when you become aware that the information has been made public, if it contains PII (for example, for GDPR, within 72 hours)

8. of Information Disclosure II

An attacker can address information through a search indexer, logger, or other such mechanism.

Threat	
	Your local machine has been a bit too zealous in indexing the files on your disk, and the index is available to other users of the machine.
CAPEC	CAPEC-643 - Identify Shared Files/Directories on System
ASVS	4.1.3 - Ensure users or services only have the necessary privileges to perform the actions they need to do.
CWE	CWE-612 - Improper Authorization of Index Containing Sensitive Information

Mitigations	
	• Purge the cache • Add exclusions to your indexer configuration

9. of Information Disclosure I

An attacker can read sensitive information in a file with bad ACLs.

Or the alternative text:

An attacker can read sensitive information in a file with permissive permissions.

Threat	
	You created a file with read or write permissions for a group or everyone, but you should be the only person able to access that file and it's a shared computer/multi-user system.
CAPEC	CAPEC-127 - Directory Indexing CAPEC-497 - File Discovery
ASVS	4.1.3 - Ensure users or services only have the necessary privileges to perform the actions they need to do. 4.3.2 - Ensure directory listing/indexing is disabled.
CWE	CWE-921 - Storage of Sensitive Data in a Mechanism without Access Control CWE-497 - Exposure of Sensitive System Information to an Unauthorized Control Sphere CWE-359 - Exposure of Private Personal Information to an Unauthorized Actor CWE-1220 - Insufficient Granularity of Access Control

Mitigations	
	• Implement a policy of least privilege • Ensure that both user and role/group access rights are set correctly • Ensure that directory browsing is disabled • Deploy a **Data Leak Prevention** (**DLP**) solution at the network level, the host level, or both

9. of Information Disclosure II

An attacker can read sensitive information in a file with bad ACLs.

Or the alternative text:

An attacker can read sensitive information in a file with permissive permissions.

Threat	
	You have an object store policy that grants access to any IAM user or service.
CAPEC	CAPEC-180 - Exploiting Incorrectly Configured Access Control Security Levels
ASVS	1.4.5 - Ensure usage of fine-grained access control such as **attribute-based access control (ABAC)**. 4.1.3 - Ensure users or services only have the necessary privileges to perform the actions they need to do.
CWE	CWE-922 - Insecure Storage of Sensitive Information CWE-921 - Storage of Sensitive Data in a Mechanism without Access Control CWE-1220 - Insufficient Granularity of Access Control

Mitigations	
	• Implement a policy of least privilege • Avoid using wildcards in policies • Use roles and add people or groups to those roles

10. of Information Disclosure

An attacker can read sensitive information in a file with no ACLs.

Or the alternative text:

An attacker can read information in files or databases with no access controls.

Threat	
	You have left an object store/bucket with public read and write access. An attacker has taken advantage of this to change the content used for your corporate identity and damage the image of your organization.
CAPEC	CAPEC-180 - Exploiting Incorrectly Configured Access Control Security Levels
ASVS	1.4.5 - Ensure usage of fine-grained access control such as ABAC. 4.1.3 - Ensure users or services only have the necessary privileges to perform the actions they need to do.
CWE	CWE-922 - Insecure Storage of Sensitive Information CWE-921 - Storage of Sensitive Data in a Mechanism without Access Control CWE-1220 - Insufficient Granularity of Access Control

Mitigations	
	• Ensure the default setting for your account is to set object stores as private on creation • Implement a policy of least privilege

Jack of Information Disclosure

An attacker can discover the fixed key being used to encrypt.

Threat	
	You've used the same key everywhere and the key has global read permissions, enabling an attacker to steal the key in one place and gain access everywhere.
CAPEC	CAPEC-180 - Exploiting Incorrectly Configured Access Control Security Levels
ASVS	4.1.3 - Ensure users or services only have the necessary privileges to perform the actions they need to do.
CWE	CWE-921 - Storage of Sensitive Data in a Mechanism without Access Control
	CWE-497 - Exposure of Sensitive System Information to an Unauthorized Control Sphere

Mitigations	
	• Like passwords, don't use the same key everywhere
	• Make sure the access control permissions for the file are set correctly
	• Maintain a key revocation list. Keys are then routinely checked against as part of the authentication/authorization process
	• Rotate keys regularly using automation, and immediately if there is a risk that one of the keys was compromised
	• Update policies and procedures defining your key usage and key rotation policy

Queen of Information Disclosure

An attacker can read the entire channel because the channel (say, HTTP or SMTP) isn't encrypted.

Threat	
	Your API doesn't require HTTPS and authentication is in cleartext, so an attacker is harvesting the credentials of your customers.
CAPEC	CAPEC-94 - **Adversary in the Middle (AiTM)** CAPEC-466 - Leveraging Active Adversary in the Middle Attacks to Bypass Same Origin Policy
ASVS	1.9.1 - Ensure you're using TLS everywhere. 9.1.1 - Ensure the TLS version can't be downgraded.
CWE	CWE-319 - Cleartext Transmission of Sensitive Information

Mitigations	
	• Ensure that you use TLS • Adopt a zero-trust culture and even encrypt local network traffic • Redirect all HTTP traffic to HTTPS

King of Information Disclosure

An attacker can read network information because there's no cryptography used.

Threat	
	You haven't architected for zero trust (an architecture to protect against bad actors that are already inside your network). For this reason, many of your internal applications use HTTP instead of HTTPS. Once inside your network, an attacker can just sit there with a packet sniffer and harvest credentials, emails, personal data, finance data, you name it. Wireshark can quite easily reconstruct the stream of packets so that you can glue a conversation back together.
CAPEC	CAPEC-94 - Adversary in the Middle (AiTM) CAPEC-157 - Sniffing Attacks CAPEC-158 - Sniffing Network Traffic
ASVS	1.9.1 - Ensure you're using TLS everywhere. 9.1.1 - Ensure the TLS version can't be downgraded.
CWE	CWE-319 - Cleartext Transmission of Sensitive Information

Mitigations	
✓	Depending on the protocol and the content, use one of the following: • Use encryption in transit • Use field or message-level encryption

Ace of Information Disclosure I

You've invented a new information disclosure attack.

Threat	
	Your staff in tech support, customer services, finance, or some other department receive a call from someone asking for information urgently. The attacker is performing a technique called pretexting and will often use pressure and urgency to rush staff into not thinking clearly.
CAPEC	CAPEC-416 - Manipulate Human Behavior CAPEC-407 - Pretexting CAPEC-412 - Pretexting via Customer Service CAPEC-415 - Pretexting via Phone
ASVS	N/A
CWE	N/A

Mitigations	
	• User education/awareness training • Implement a threat intelligence program to keep staff informed of any new or active threats they should be aware of • Notify staff of any incidents or threat actor attempts to compromise the security of the organization • Update your security program with any additional controls needed to protect against these threats

Ace of Information Disclosure II

You've invented a new information disclosure attack.

Threat	
	Your HTTP response headers contain information about your host environment. An attacker can call your services or web applications with a HEAD request, gather information about the server software being used, and then determine whether that version is vulnerable. This is called fingerprinting, and there are many variants of this sort of probing reconnaissance.
CAPEC	CAPEC-224 - Fingerprinting CAPEC-170 - Web Application Fingerprinting
ASVS	14.3.3 - Filter HTTP headers that disclose security-sensitive system information such as software/OS versions.
CWE	CWE-497 - Exposure of Sensitive System Information to an Unauthorized Control Sphere CWE-1230 - Exposure of Sensitive Information through Metadata CWE-212 - Improper Removal of Sensitive Information Before Storage or Transfer

Mitigations	
	• Disable any information being returned in headers that can give an attacker an advantage and that isn't strictly necessary for the running of your applications • If the software doesn't have configuration options to remove this data, use filters on the firewall to remove it

E of Information Disclosure

Personal data is being sent over a plaintext connection or email.

Threat	
	You're communicating the personal information of your customers via some means that isn't encrypted so an attacker who is capturing your network traffic as it passes over the internet can read their details.
CAPEC	CAPEC-94 - Adversary in the Middle (AiTM) CAPEC-157 - Sniffing Attacks CAPEC-158 - Sniffing Network Traffic
ASVS	1.9.1 - Ensure you're using TLS everywhere. 9.1.1 - Ensure the TLS version can't be downgraded. 9.2.2 - Ensure TLS is also used for monitoring and management interfaces and can't be downgraded.
CWE	CWE-319 - Cleartext Transmission of Sensitive Information

Mitigations	
	• For email, S/MIME would protect the content of the emails • Use TLS for data in transit • Consider encrypting the data at the file level • Ensure that your company policies and procedures contain guidance on the correct handling of customer data and its communication • Ensure that contracts with any third parties you use for data processing contain guidance and penalty clauses for non-compliance

F of Information Disclosure

Personal data is being saved on unencrypted media.

Threat	
	You are saving personal data on a USB key or your computer without having either full disk encryption or encrypting the data directly, so if a thief were to steal your computer, they could read all of your personal or your company's data.
CAPEC	CAPEC-507 - Physical Theft
ASVS	1.8.2 - Ensure the level of protection (confidentiality, integrity, and availability) matches the security and privacy classification of the data. 6.1.1 - Ensure all PII is encrypted at rest in line with the **General Data Protection Regulation (GDPR)** requirements. 6.1.2 - Ensure all medical data is encrypted at rest in line with the **Health Insurance Portability and Accountability Act (HIPAA)** requirements. 6.1.3 - Ensure all financial data is encrypted at rest in line with **Payment Card Industry Data Security Standard (PCI-DSS)** requirements.
CWE	CWE-312 - Cleartext Storage of Sensitive Information

Mitigations	
	• Use full disk encryption (this is transparent though, meaning that if the disk has been mounted, it can be read by anyone accessing the machine) • Use file-level encryption (only people with the key can read the data) • Lock your computer before leaving it unattended in the office • Disconnect your devices and store them safely • Log out of your computer and switch it off when it is not in use • Disable USB for connection of storage devices in the BIOS or at a policy level

Summary

You've now covered the threat types described on the cards from the Information Disclosure suit in the Elevation of Privilege card deck with the addition of two cards from the T.R.I.M. extension for the game. These threats detailed flaws relating to brute forcing, reconnaissance, missing cryptography both in transit and at rest, being too descriptive in messages, social engineering, file permissions, and access control.

You should now understand why sensitive data should be encrypted wherever possible, how sharing too much information (no matter how innocuous it may seem) can be harmful, and why you should always set the strongest permissions by default and then relax them as needed. Having a greater awareness of the threats in this category and having discussed how to mitigate them, be confident that you are making it as difficult as possible for your adversary to get at the data.

In the next chapter, we will cover the fifth S.T.R.I.D.E. category of threats or the fifth S.T.R.I.D.E. category: Denial of Service.

6
Denial of Service

We perform a **denial of service** attack when we stop something from being able to do its job. We can do this in a number of ways; it might be that you cause the system to crash, cause it to be unreachable, or stop it from performing some task. The system being attacked could be either a service or a client. We should also note that a client doesn't necessarily mean a browser but could also be a service talking to another service.

Figure 8.1: Service is being denied to the devices because of an attack

In this chapter, we'll look at several denial of service threats. It is important to keep in mind that denial of service is not only caused by attackers, but it can also be human error, file corruption, bad planning, or even an outage of a service your application depends upon. As in previous chapters, I'll give you references from CAPEC, ASVS, and CWE with each example where you can get more information. I will also suggest what mitigations and controls you can put in place to protect against the threat or at least reduce the risk.

By the end of the chapter, you'll have a better understanding and awareness of the threats that can cause your application to be unavailable or unusable and be able to recognize them in your system architecture and designs.

2. of Denial of Service I

An attacker can make your authentication system unusable or unavailable.

Threat	
	Your system locks users out after a number of failed login attempts, an attacker could enumerate users by deliberately sending invalid passwords to lock out all the users.
CAPEC	CAPEC-2 – Inducing account lockout
ASVS	2.2.1 – Verify the effectiveness of the authentication controls
CWE	CWE-307 – Improper restriction of excessive authentication attempts

Mitigations	
	• If repeated failed login attempts come from the same address for multiple accounts, block the address and not the user account

2. of Denial of Service II

An attacker can make your authentication system unusable or unavailable.

Threat	
	You use single sign-on and can no longer reach your **identity provider (IdP)** because an attacker is using **Address Resolution Protocol (ARP)** or **Domain Name System (DNS)** poisoning to make your IdP, or login server. By forcing your ARP cache for a given IP to point to a different **medium access control address (MAC)**, a unique value given to each network interface, they can stop you from being able to reach your identity provider IdP because your system no longer knows how to get to its destination.
CAPEC	CAPEC-589 – DNS Blocking
	CAPEC-590 – IP address blocking
	CAPEC-603 – Blockage
	CAPEC-607 – Obstruction
ASVS	N/A
CWE	CWE-940 – Improper verification of the source of a communication channel
	CWE-350 – Reliance on reverse DNS resolution for a security-critical action

Mitigations	
	• DNS poisoning can be avoided using DNSSEC, a security extension to the DNS protocol where responses to queries are digitally signed and can therefore be verified
	• Use network segmentation to reduce the scope of ARP poisoning
	• Use static ARP entries
	• Use packet filtering firewalls that alert when packets from the same IP are found with different MACs or when MACs associated with an IP suddenly change or vice versa
	• ARP poisoning in some cloud platforms has been made much more difficult through the use of machine identity, so it may be an avenue to explore

3. of Denial of Service I

An attacker can make a client unavailable or unusable but the problem goes away when the attacker stops (client, authenticated, temporary).

Threat	
	You don't support simultaneous sessions, or you have users who can remove connected devices that should no longer be used. An attacker is connecting with a legitimate user's credentials and taking advantage of this to invalidate their session repeatedly.
CAPEC	CAPEC-74 – Manipulating state
ASVS	3.3.3 – Ensure a change of password terminates active sessions
	3.3.5 – Ensure users can see and terminate an active session
CWE	CWE-1018 – Manage user sessions

Mitigations	
	• Implement **multifactor authentication** (MFA) so the attacker cannot log in with the legitimate user's credentials
	• Make reset password functionality invalidate all sessions so the legitimate user is once again the only one with access to the account

3. of Denial of Service II

An attacker can make a client unavailable or unusable but the problem goes away when the attacker stops (client, authenticated, temporary).

Threat	
	You are using a third-party JavaScript library and an attacker changes the source code, a supply chain attack, to inject JavaScript code into your site, obliging clients to make multiple requests to his site which performs a slow read attack, forcing the client to read one byte at a time.
CAPEC	CAPEC-446 – Malicious logic insertion into the product via the inclusion of a third-party component
ASVS	10.1.1 – Ensure you're scanning your code for vulnerabilities as part of your secure development process
CWE	CWE-74 – Improper neutralization of special elements in output used by a downstream component (injection)

Mitigations	
	• Use **cross-origin resource sharing** (**CORS**) policies to block any requests to any unexpected locations. These are headers that instruct the browser to only get content from certain trusted sources.
	• Implement SAST scanning as part of your secure development process.

3. of Denial of Service (alternative 2022 deck)

Alternative 2022 deck

On January 20, 2022, Adam Shostack posted an article on his blog suggesting some new cards and edits or alternatives for existing cards. You can find his blog post here: `https://shostack.org/blog/elevation-of-privilege-2022/`.

An attacker can drain our easily replaceable battery (battery, temporary)

Threat	
	You are not rate limiting the number of simultaneous complex queries a user can perform. By causing excessive use of the processor in calculation or repeated tasks, an attacker could cause energy consumption to peak unexpectedly and therefore drain the battery you are using as the accumulator in a solar-powered installation.
CAPEC	CAPEC-124 – Shared resource manipulation
	CAPEC-130 – Excessive allocation
ASVS	N/A
CWE	CWE-399 – Resource management errors
	CWE-400 – Uncontrolled resource consumption

Mitigations	
	• Ensure you can switch over from one battery to another without having downtime
	• Load test the system to determine at what threshold this could occur
	• Rate limit and throttle usage from any user exceeding predefined usage criteria based on load tests
	• Alert when processes are grabbing too much of any resource

4. of Denial of Service

An attacker can make a server unavailable or unusable but the problem goes away when the attacker stops (server, authenticated, temporary).

Threat	
	This isn't a persistent attack, which should immediately suggest the attacker is saturating or blocking one of the following resources: • Network • CPU • Memory Once they stop attacking, these will gradually be freed up, therefore things will go back to normal. This could be by launching multiple long-running processes in parallel, or it might be by sending many requests containing large amounts of data simultaneously. Either of these could be intentional or simply ingenuity on the part of a user.
CAPEC	CAPEC-131 – Resource leak exposure CAPEC-124 – Shared resource manipulation
ASVS	N/A
CWE	CWE-399 – Resource management errors CWE-400 – Uncontrolled resource consumption

Mitigations	
	• Rate limit users based on the number of requests • Rate limit users based on bandwidth consumption • Rate limit users based on the length of time processing requests

4. of Denial of Service (alternative 2022 deck)

An attacker can drain a battery that's hard to replace (sealed in a phone, an implanted medical device, or in a hard-to-reach location) (battery, persist).

Threat	
	You've decided to add remote access to a new pacemaker you are developing without considering all the possible risks this could introduce. Some medical devices such as pacemakers are configurable remotely via radio and need to have low energy consumption because they have either rechargeable or long-life batteries. An attacker is continually sending meaningless data to the device though, and it must decide whether that data is meaningful, which wastes the clock cycles of the microprocessor and consequently energy.
CAPEC	CAPEC-262 – Manipulate system resources CAPEC-130 – Excessive allocation
ASVS	N/A
CWE	CWE-399 – Resource management errors CWE-400 – Uncontrolled resource consumption

Mitigations	
	• In the case of a pacemaker, implement a means of activating and deactivating the radio communications of the device • Use a magic number at the start of requests being sent so that you can immediately drop packets if the first few bytes are invalid • Solicit requirements in the design phase by creating abuse cases and security use cases • Use threat intelligence to understand what attacks your product might be susceptible to by adding new features (e.g., remote access) • Implement a means of charging the battery, perhaps by induction to keep it constantly topped up

5. of Denial of Service I

An attacker can make a client unavailable or unusable without ever authenticating, but the problem goes away when the attacker stops (client, anonymous, temporary).

Threat	
	The attacker is sitting between the client and the server, performing deep packet inspection and selectively dropping packets to block access to the endpoint.
CAPEC	CAPEC-590 – IP address blocking
ASVS	N/A
CWE	N/A

Mitigations	
	• Alert if packet loss exceeds a predefined threshold • Use network diagnostic tools such as traceroute to determine the path being taken • Automate rerouting of packets to avoid a path causing significant packet loss • Use backup IPs so that if one is blocked, it may be possible to reach an alternative

5. of Denial of Service II

An attacker can make a client unavailable or unusable without ever authenticating, but the problem goes away when the attacker stops (client, anonymous, temporary).

Threat	
	You rate limit users of your system, so an attacker has stolen a user's credentials and is exceeding usage limits on your service, causing rate limiting to take place that is affecting the legitimate user.
CAPEC	CAPEC-2 – Inducing account lockout
ASVS	N/A
CWE	CWE-399 – Resource management errors

Mitigations	
	• If this interaction is human-in-the-loop, require MFA for authentication and ask them to re-authenticate
	• If this is an API, and you are using OAuth, then use the short time to live values for access tokens, forcing clients to use their refresh token

5. of Denial of Service (alternative 2022 deck)

An attacker can spend our cloud budget (budget, persist).

Threat	
	You are using elastic compute, so the attacker creates an excessive load on the system, causing it to keep scaling until the costs for the usage exceed any cap you have put on the budget for the current month.
CAPEC	CAPEC-130 – Excessive allocation
ASVS	N/A
CWE	CWE-400 – Uncontrolled resource consumption

Mitigations	
	• Limit scaling within your budget
	• Create alerts when costs exceed different levels
	• Use a WAF to protect against DOS attacks and other common attacks
	• Throttle connections from a single source

6. of Denial of Service I

An attacker can make a server unavailable or unusable without ever authenticating, but the problem goes away when the attacker stops (server, anonymous, temporary).

Threat	
	You are temporarily locking users out of their accounts after three failed login attempts to protect against brute force attacks. An attacker is taking advantage of this security control and making deliberate repeated failed logins to cause account lockouts.
CAPEC	CAPEC-2 – Inducing account lockout
ASVS	N/A
CWE	CWE-645 – Overly restrictive account lockout mechanism

Mitigations	
	• Ensure temporary lockouts for failed authentication attempts are restricted not only by the user but also by the source

6. of Denial of Service II

An attacker can make a server unavailable or unusable without ever authenticating, but the problem goes away when the attacker stops (server, anonymous, temporary).

Threat	
	You don't have a firewall in front of your server, so an attacker uses network flooding, such as TCP flood, UDP flood, ICMP flood, HTTP flood, or SSL flood to render your server unavailable.
CAPEC	CAPEC-482 – TCP flood CAPEC-486 – UDP flood CAPEC-487 – ICMP flood CAPEC-488 – HTTP flood CAPEC-489 – SSL flood
ASVS	N/A
CWE	N/A

Mitigations	
	• Ensure your WAF is configured to block this kind of attack

7. of Denial of Service

An attacker can make a client unavailable or unusable and the problem persists after the attacker goes away (client, authenticated, persistent).

Threat	
	An attacker sends a phishing email to a user who logs in to a fake version of your site, and the attacker then logs into your site with the credentials they gathered and changes the password, locking out the legitimate user.
CAPEC	CAPEC-98 – Phishing CAPEC-163 – Spear phishing
ASVS	2.2.4 – Ensure the use of MFA
CWE	CWE-308 – Use of single-factor authentication

Mitigations	
	• Require MFA • Detect whether the user is connecting from a new location or a new device • Alert the user of password changes with a link to flag this as illegitimate if not performed by them

8. of Denial of Service

An attacker can make a server unavailable or unusable and the problem persists after the attacker goes away (server, authenticated, persistent).

Threat	
	You are not validating or sanitizing untrusted user input before querying the database. Using SQL injection, an attacker deletes all the users of the system from the database.
CAPEC	CAPEC-66 – SQL injection
ASVS	5.3.4 – Ensure you're using parameterized queries
CWE	CWE-89 – Improper neutralization of special elements used in an SQL command (SQL injection)

Mitigations	
	• Use parametrized queries; most programming languages offer support for parameterized queries and separating the query logic from the data will render any values being passed as secure
	• Validate and sanitize data coming from untrusted sources

9. of Denial of Service

An attacker can make a client unavailable or unusable without ever authenticating, and the problem persists after the attacker goes away (client, anonymous, persistent).

Threat	
	You are not sanitizing untrusted input before writing it to your database, and as a consequence, an attacker has managed to store a DOM-based **cross-site scripting (XSS)** attack in your database that causes an infinite loop or recursion in the JavaScript on your site, causing your clients' CPUs to max out.
CAPEC	CAPEC-592 – Stored XSS CAPEC-588 – DOM-based XSS
ASVS	5.3.3 – Ensure there is sanitization of XSS not just for input but also for output
CWE	CWE-80 – Improper neutralization of script-related HTML tags in a web page (basic XSS)

Mitigations	
	• Always sanitize input from untrusted sources • Escape output within the context it will be used; for example, if data is to be used in JavaScript, ensure that the data has escaped so that any script it contains will not be executed

10. of Denial of Service

An attacker can make a server unavailable or unusable without ever authenticating, and the problem persists after the attacker goes away (server, anonymous, persistent).

Threat	
	You have a file upload endpoint that doesn't limit the file size, so an attacker sends enough data to fill the storage on the server and stop it from being able to further process requests.
CAPEC	CAPEC-231 – Oversized serialized data payloads CAPEC-572 – artificially inflate file sizes
ASVS	5.2.2 – Ensure data is sanitized
CWE	CWE-400 – Uncontrolled resource consumption

Mitigations	
	• Limit the file size of uploads • Limit the number of uploads from a single source • Validate files being uploaded and only store valid files

Jack of Denial of Service I

An attacker can cause the logging subsystem to stop working.

Threat	
	Your log configuration files are not read-only, so an attacker can modify the log level being used by the system from info to fatal so that none of the logs defined in the code are triggered.
CAPEC	CAPEC-571 – Block logging to the central repository
ASVS	7.1.3 – Ensure the application logs security-related events 7.1.4 – Ensure log entries contain what's necessary for forensics
CWE	CWE-778 – Insufficient logging

Mitigations
• Ensure that strong ACLs are used on all configuration files • Add a heartbeat to your logs and create alerts if no longer present

Jack of Denial of Service II

An attacker can cause the logging subsystem to stop working.

Threat	
	Your log files are owned by the technical user running the application, so an attacker can take advantage of a remote code execution vulnerability in your system to run code that changes the file permissions on the log files. Now, the application can no longer write logs to the files.
CAPEC	CAPEC-571 – Block logging to the central repository
ASVS	7.3.3 – Ensure privileges are set correctly for log access
CWE	CWE-778 – Insufficient logging

Mitigations	
	• Make a privileged user the owner of the file, so that the service/process user cannot change permissions that would result in write access being denied to the logging subsystem

Queen of Denial of Service I

An attacker can amplify a denial of service attack through this component with amplification on the order of 10:1.

Threat	
	A part of your system doesn't restrict the size of the data it can process, and you also have an endpoint that allows you to perform batch processing with this already vulnerable part of the system.
CAPEC	CAPEC-572 – Artificially inflate file sizes CAPEC-231 – Oversized serialized data payloads
ASVS	12.1.1 – Ensure files won't fill your disk 12.1.2 – Ensure extracted archives won't fill your disk
CWE	CWE-770 – Allocation of resources without limits or throttling

Mitigations	
	• Ensure endpoints that perform batch processing don't accept large files • Limit the number of concurrent batch jobs allowed per user

Queen of Denial of Service II

An attacker can amplify a denial of service attack through this component with amplification on the order of 10:1.

Threat	
	You have a search feature that allows multiple search terms at the same time. For each search term, a separate search is performed internally. An attacker has cleverly crafted several search terms that will take a long time to process and has then launched multiple searches at the same time.
CAPEC	CAPEC-130 – Excessive allocation CAPEC-490 – Amplification
ASVS	13.4.1 – Ensure checks are performed to protect against exponential or uncontrolled
CWE	CWE-674 – Uncontrolled recursion CWE-776 – Improper restriction of recursive entity references in DTDs (XML entity expansion) CWE-400 – Uncontrolled resource consumption CWE-770 – Allocation of resources without limits or throttling

Mitigations	
	• Limit the number of concurrent searches allowed per user • In this example, consider limiting the number of search terms if this spawns multiple searches

King of Denial of Service

An attacker can amplify a denial of service attack through this component with amplification on the order of 100:1.

Threat	
	Your system accepts multiple queries that are executed simultaneously, and each of those queries is subsequently broken down into multiple smaller queries. This brings about an exponential number of queries being executed from a single request.
CAPEC	CAPEC-130 – Excessive allocation CAPEC-490 – Amplification
ASVS	13.4.1 – Ensure checks are performed to protect against exponential or uncontrolled recursive querying
CWE	CWE-674 – Uncontrolled recursion CWE-776 – Improper restriction of recursive entity references in DTDs (XML entity expansion) CWE-400 – Uncontrolled resource consumption CWE-770 – Allocation of resources without limits or throttling

Mitigations
<ul><li>Limit the complexity of queries allowed for batch processing</li><li>Limit the number of concurrent complex queries allowed per user</li><li>Timeout long-running queries</li></ul>

Ace of Denial of Service

You've invented a new denial of service attack.

Threat	
	Your entire organization is connected over Wi-Fi and an attacker outside the building is jamming the Wi-Fi signal, thereby disrupting all communication within the organization.
CAPEC	CAPEC-604 – Wi-Fi Jamming CAPEC-601 – Jamming CAPEC-607 – Obstruction
ASVS	N/A
CWE	N/A

Mitigations	
	• Consider shielding the building but this could interrupt other forms of communication • Consider cabling critical systems

E of Denial of Service

The availability of certain personal data is a life-or-death matter, and our system is not as reliable as it should be.

Threat	
	Your system is a patient monitoring system in a hospital, and it is used to monitor patients whose vital signs are critical. Each monitor is linked to the nursing station centrally, so they are notified in real time if a patient requires assistance. The communication is over Wi-Fi and the building structure is causing problems with signal strength and notifications are not being received.
CAPEC	CAPEC-604 – Wi-Fi Jamming
	CAPEC-607 – Obstruction
ASVS	N/A
CWE	N/A

Mitigations	
	• If real-time communication is required, consider using a wired solution
	• If a wired solution is not feasible, ensure you have sufficient signal repeaters or access points
	• Ensure the system acknowledges receipt of notifications and implement a retry strategy
	• Use audible and visual alerts local to the patient to attract attention

Summary

You've now covered the threat types described on the cards from the Denial of Service suit in the Elevation of Privilege card deck. You've seen how an attacker doesn't necessarily need to attack you directly for your application to be affected. You've seen how the design of features meant to protect your users can be used against you and how exponential growth can be devastating to resources.

You should now understand the following:

- How a security control, if implemented incorrectly, can itself become a threat
- You should be able to consider the wider interaction of your application
- The application should not allow a single process or user to overwhelm system resources
- If the system is required to run complex tasks simultaneously and remain performant, then it should be able to scale to cater to demand

In the next chapter, we will cover the Elevation of Privilege category of threats from **STRIDE** and the Elevation of Privilege game; this is the trump deck and you will soon discover its importance.

Elevation of Privilege

Elevation of Privilege is when a user can execute something at a level of privilege superior to the one they should have, meaning they can acquire additional access rights or permissions or even take on the role of a different user. The ultimate goal is to have complete control of a system. This is called **vertical privilege escalation**, and there is also another type called **horizontal privilege escalation**. In horizontal privilege escalation, instead of taking on a role with a higher level of privilege, you take on the role of another user with the same level of permission as you but you can now access systems, functions, and data they can access. For example, you are logged in to your bank and, through horizontal privilege escalation, you take on the role of another user and have access to their bank account.

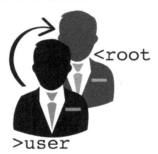

Figure 7.1: A user elevating their privileges to become the root user

In this chapter, we will cover the threats described on the cards from the Elevation of Privilege suit in the Elevation of Privilege card deck. We'll look at several Elevation of Privilege threats, which, as the name suggests, allow an attacker to get more privileges with that power. As in previous chapters, I'll give you references from CAPEC, ASVS, and CWE with each example where you can get more information. I will also suggest what mitigations and controls you can put in place to protect against the threat or at least reduce the risk.

By the end of the chapter, you'll have a better understanding and awareness of the threats that allow an attacker to gain more privileges and be able to recognize them in your system architecture and designs.

Before we get started, let me introduce a few concepts.

Some important concepts

A **supply chain attack** is when a bad actor attacks one of your suppliers as a means to then attack you. They know that there is a trust relationship between you and your supplier and that if they can compromise your supplier in some way, then they can abuse that trust. This could be by inserting malware into a software update for a product you have purchased.

When we talk about **access control**, we're referring to the three As – **authentication**, **authorization**, and **accounting**. There are a number of different access control models but I won't go into detail here; **mandatory access control** (**MAC**), **discretionary access control** (**DAC**), **role-based access control** (**RBAC**), and **attribute-based access control** (**ABAC**) are just a few of them.

Access control lists (**ACLs**) define what users, groups, or roles have sufficient access permissions to be able to read or write to a file, for example.

We'll also talk about **multifactor authentication** (**MFA**) – the different factors refer to something you have, something you know, and something you are:

- You have, for example, a debit card and you know the PIN (something you have and something you know)
- You have a username and you unlock your computer with your fingerprint (something you know and something you are)

Multifactor is when you have two or more of these, adding an additional level of security.

Injection attacks or **injection** is when the computer accepts input from a user and it then processes that input incorrectly, for example, executing it as though it were a program. This can be avoided by rejecting input that contains any characters you don't expect (validation), removing any characters you don't expect (sanitization), or converting any characters you don't expect into a form that can't do any harm (escaping), which is also a form of sanitization.

With the basics covered, let's shift our focus to the cards.

2. of Elevation of Privilege (2022 deck)

An attacker has compromised a key technology supplier.

Threat	
	The company that writes your order processing system has been hacked and the attackers have written a backdoor into the latest version and you have configured automatic updates. So, they can now steal/exfiltrate all your customer data.
CAPEC	CAPEC-523 – Malicious software implanted CAPEC-511 – Infiltration of the software development environment CAPEC-657 – Malicious automated software update via spoofing
ASVS	10.2.1 – Check for application phoning home and harvesting of data 10.2.3 – Check source for backdoors and other malicious code 10.3.1 – Check that automatic updates are signed and performed over a secure channel
CWE	CWE-494 – Download the code without an integrity check

Mitigations	
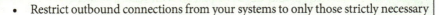	• Restrict outbound connections from your systems to only those strictly necessary • Install a **host-based intrusion detection system** (**HIDS**) and a **network intrusion detection system** (**NIDS**) to detect any anomalous behavior • Always vet your suppliers verifying their track record • In your vendor contracts, stipulate what you require of them for developing and operating solutions in a secure manner

3. of Elevation of Privilege (2022 deck) I

An attacker can access the cloud service that manages your devices.

Threat	
	Your burglar alarm and video surveillance are connected to the cloud but the cloud you are using isn't securely configured, so a bad actor can connect and watch when you are at home from a remote location and disable your alarm so that they can break in.
CAPEC	CAPEC-1 – Accessing functionality not properly constrained by ACLs CAPEC-565 – Password spraying CAPEC-180 – Exploiting incorrectly configured access control security levels
ASVS	4.3.1 – Ensure usage of MFA
CWE	CWE-1220 – Insufficient granularity of access control

Mitigations	
	• Ensure you have the correct security configuration settings in your cloud environment • Keep your devices updated with security patches • Add alerting to your cloud account so that you know if someone accesses it without your consent • Use MFA to protect user accounts

3. of Elevation of Privilege (2022 deck) II

An attacker can access the cloud service which manages your devices.

Threat	
	Your cloud console admin account isn't secure, so an attacker can spin up new instances for crypto mining, which are charged back to you.
CAPEC	CAPEC-1 – Accessing functionality not properly constrained by ACLs CAPEC-565 – Password spraying CAPEC-180 – Exploiting incorrectly configured access control security levels
ASVS	4.3.1 – Ensure usage of MFA
CWE	CWE-1220 – Insufficient granularity of access control

Mitigations	
	• Ensure you have the correct security configuration settings in your cloud environment • Add alerting to your cloud account so that you know if someone accesses it without your consent • Use MFA to protect user accounts • Use separation of duties and separate your billing/finance and administration accounts

4. of Elevation of Privilege (2022 deck)

An attacker can escape from a container or other sandbox.

Threat	
	You are using the public cloud or are running your application on a multi-tenant environment (a physical computer shared between multiple organizations, running multiple **virtual machines** (**VMs**) or software for each of them); an attacker is running in another VM or container on the same environment and is able to break out of their hypervisor or container and attack other tenants.
CAPEC	CAPEC-480 – Escaping virtualization CAPEC-233 – Privilege escalation
ASVS	N/A
CWE	CWE-668 – Exposure of resources to the wrong sphere

Mitigations	
	• Operate a zero-trust model • Ensure you are encrypting your data at rest • Harden your container • Where possible, use dedicated hardware (a single physical computer running only your organization's VMs or software) if your system contains mission-critical data • Patch and harden the container host • Remove all unused software • Avoid running containers as root

5. of Elevation of Privilege I

An attacker can force data through different validation paths, which give different results.

Threat	
	You are validating input at different places in your application and not using centralized functions to perform the checks, so there may be differences in the implementation and other pieces of code may be making certain assumptions about those checks.
CAPEC	CAPEC-80 – Using UTF-8 encoding to bypass validation logic CAPEC-71 – Using Unicode encoding to bypass validation logic CAPEC-64 – Using slashes and URL encoding combined to bypass validation logic
ASVS	1.1.6 – Ensure the use of a security kernel
CWE	CWE-20 – Improper input validation

Mitigations	
	• Centralize functions that perform validation and sanitization, as this reduces effort if you need to update a pattern • Always check that any assumptions you have made are correct

5. of Elevation of Privilege II

An attacker can force data through different validation paths, which give different results.

Threat	
	Your site is multilingual and implementations for each language may differ slightly. An attacker can change the default language in their browser, causing a change in flow so that they can take advantage of some missing validation.
CAPEC	CAPEC-554 – Functionality bypass CAPEC-140 – Bypassing of intermediate forms in multiple-form sets CAPEC-29 – Leveraging **time-of-check and time-of-use (TOCTOU)** race conditions
ASVS	1.11.2 – Verify thread safety 1.11.3 – Check for TOCTOU issues 11.1.1 – Verify application flow is enforced
CWE	N/A

Mitigations	
	• Ensure validation functionality is centralized and shared between the different implementations • Always validate and sanitize untrusted input • Verify whether assumptions are correct and are documented

6. of Elevation of Privilege

An attacker could take advantage of .NET permissions you ask for but don't use.

Or the alternative text:

An attacker could take advantage of permissions you set but don't use.

Threat	
	You've given the service user that launches your application permission to read and write to the file system, but they only need to have read access; consequently, because of a flaw in your application, an attacker may overwrite critical system files.
CAPEC	CAPEC-180 – Exploiting incorrectly configured access control security levels
ASVS	10.2.2 – Check that only required permissions are assigned
CWE	CWE-276 – Incorrect default permissions

Mitigations	
	• Always use the principle of least privilege • Audit permissions regularly to ensure any permissions no longer needed are removed

7. of Elevation of Privilege

An attacker can provide a pointer across a trust boundary, rather than data that can be validated.

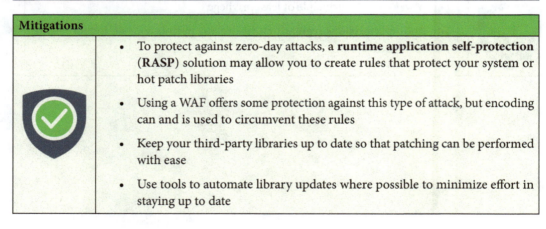

Threat	
	The **Log4Shell** (`https://en.wikipedia.org/wiki/Log4Shell`) vulnerability was a classic example of this, where an attacker could pass a reference that pointed back to themselves, which was interpreted by the `log4j` library using the **Java Naming and Directory Interface** (**JNDI**). Through this, attackers were able to create reverse shells, allowing them to compromise your systems.
CAPEC	CAPEC-253 – Remote code inclusion
ASVS	N/A
CWE	CWE-117 – Improper output neutralization for logs
	CWE-1395 – Dependency on vulnerable third-party component
	CWE-917 – Improper neutralization of special elements used in an expression language statement (expression language injection)

Mitigations	
	• To protect against zero-day attacks, a **runtime application self-protection** (**RASP**) solution may allow you to create rules that protect your system or hot patch libraries
	• Using a WAF offers some protection against this type of attack, but encoding can and is used to circumvent these rules
	• Keep your third-party libraries up to date so that patching can be performed with ease
	• Use tools to automate library updates where possible to minimize effort in staying up to date

8. of Elevation of Privilege

An attacker can enter data that is checked while still under the attacker's control and used later on the other side of the trust boundary.

Threat	
	You are validating and/or sanitizing user input on the client side but an attacker can circumvent this using developer tools in the browser, writing their own scripts, or proxying calls to the backend. For example, you validate authentication information to avoid **Structured Query Language (SQL)** injection but aren't sanitizing or using parametrized queries on the backend.
CAPEC	CAPEC-2 – Exploiting trust in the client
	CAPEC-77 – Manipulating user-controlled variables
ASVS	1.5.3 – Ensure validation is performed where it can't be tampered with
CWE	CWE-179 – Incorrect behavior order – early validation
	CWE-807 – Reliance on untrusted inputs in a security decision

Mitigations	
	• Validate and sanitize on the frontend to avoid round-tripping
	• Validate and sanitize on the backend where the attacker doesn't have control as well as the frontend
	• For SQL injection, use parametrized queries
	• For other types of injection, potentially use allow listing

9. of Elevation of Privilege

There's no reasonable way for callers to figure out what validation of tainted data you perform before passing it to them.

Threat	
	You've not documented what you are and are not validating/sanitizing so others may be making incorrect assumptions that you are handling security on your side, and you are assuming they are securing things on theirs. This leads to neither side securing things properly.
CAPEC	N/A
ASVS	1.1.4 – Verify the system and IO functions have been documented properly
CWE	CWE-1111 – Incomplete I/O documentation

Mitigations	
	• Document functions that deal with tainted data so that users of the function know how they should behave • If you are using functions of internally developed (second-party) libraries or third-party libraries, implement handling of tainted data on your side as well to be secure by default

10. of Elevation of Privilege

There's no reasonable way for a caller to figure out what security assumptions you make.

Threat	
	You haven't documented what you expect your consumers to sanitize before sending it to you, so they are sending you unchecked/unsanitized data, leaving them or you subject to risk if you don't handle it.
CAPEC	CAPEC-153 – Input data manipulation
ASVS	5.1.3 – Validate all untrusted data and use allow lists where possible
CWE	CWE-1173 – Improper use of the validation framework CWE-183 – Permissive list of allowed inputs CWE-184 – Incomplete list of disallowed inputs CWE CATEGORY – Data neutralization issues

Mitigations	
	• Document your assumptions and expectations • Always sanitize untrusted input, even if the expectation is that the data was sanitized somewhere else

Jack of Elevation of Privilege

An attacker can reflect input back to a user, such as in **cross-site scripting** (XSS).

Threat	
	You're not checking data entered in web forms, so when an attacker puts script or HTML into a form field, that data is sent to the server, and you then include this data in the response, which, when returned to the browser, is executed or rendered.
CAPEC	CAPEC-63 – XSS CAPEC-591 – Reflected XSS CAPEC-588 – DOM-based XSS
ASVS	5.1.3 – Validate all untrusted data and use allow lists where possible 5.3.3. – Make sure you're sanitizing or encoding your output as well as your input
CWE	CWE-79 – Improper neutralization of input during web page generation (XSS)

Mitigations	
	<ul><li>Never trust user input; always validate untrusted data</li><li>Where possible, use allow listing of characters</li><li>Sanitize input, encoding the characters so that if they are returned, they are innocuous</li><li>Ensure you are using industry-standard, proven libraries to perform the sanitization</li></ul>

Queen of Elevation of Privilege I

You include user-generated content within your page, possibly including the content of random URLs.

Threat	
	Perhaps you offer a portal where users can showcase their web templates, graphic templates, or office templates and allow them to link directly to the files on their site. An attacker could make use of the trust users have in your site to deliver their malware.
CAPEC	CAPEC-17 – Using malicious files CAPEC-23 – File content injection
ASVS	1.12.2 – Ensure files are vetted so you don't serve up malicious code to other users
CWE	CWE-434 – Unrestricted upload of file with dangerous type

Mitigations	
	• Only allow external links from trusted sources • There are services available to check that links are safe before sharing them; for example, Mimecast URL Protection, NordVPN, Bitdefender, and VirusTotal • Quarantine links until checked • Don't share user content coming from outside of your site

Queen of Elevation of Privilege II

You include user-generated content within your page, possibly including the content of random URLs.

Threat	
	Your site allows users to add comments and they can insert code into the comments, which is stored directly in your database without sanitization. Subsequently, you are including it in a rendered page again without sanitization.
CAPEC	CAPEC-592 – Stored XSS
ASVS	5.1.3 – Validate all untrusted data and use allow lists where possible 5.2.1 – Ensure all untrusted input is sanitized 14.4.3 – Ensure you have applied Content Security Policies
CWE	CWE-183 – Permissive list of allowed inputs CWE-184 – Incomplete list of disallowed inputs

Mitigations	
	• Sanitize all untrusted data both on input and output to protect both your systems and those of your users • Encode user-supplied data before rendering to ensure any code is neutralized and rendered as text

King of Elevation of Privilege I

An attacker can inject a command that the system will run at a higher privilege level.

Threat	
	Your application server is launched with an admin user account instead of with a service account and an attacker manages to inject a command that is run in the **operating system (OS)** as the admin user.
CAPEC	CAPEC-233 – Privilege escalation CAPEC-69 – Target programs with elevated privileges
ASVS	1.2.1 – Ensure you're not using service accounts with only the permissions they need 2.10.2 – Ensure service to service auth is not performed as root
CWE	CWE-250 – Execution with unnecessary privileges CWE-78 – Improper neutralization of special elements used in an OS command (OS command injection)

Mitigations	
	• Never run services or web applications as the root/admin user

King of Elevation of Privilege II

An attacker can inject a command that the system will run at a higher privilege level.

Threat	
	Your application connects to a database, not in the context of the current user but as a shared database user with a high level of permissions. This is to permit different types of users to perform different tasks. You have defined what queries can be run by each type of user, but when a user manages to inject a query that performs an action, they should not have permission to do, it is executed as the shared user with elevated permissions.
CAPEC	CAPEC-7 – Blind SQL injection

CAPEC-69 – Target programs with elevated privileges |
| ASVS | 1.2.2 – Ensure you're adhering to the principle of least privilege |
| CWE | CWE-89 – Improper neutralization of special elements used in an SQL command (SQL injection) |

Mitigations	
	• Use parametrized queries to avoid SQL injection
• Execute queries in the context of the user, with single sign-on through to the DB where possible

Ace of Elevation of Privilege

You've invented a new Elevation of Privilege attack.

Threat	
	You have both an admin interface and a user interface to your web application and on login, you redirect the users to the relevant area of your site. However, you have not implemented object-level access control, so, if a user knows the URL of an administration page, the system will let them access it.
CAPEC	CAPEC-1 – Accessing functionality not properly constrained by ACLs CAPEC-180 – Exploiting incorrectly configured access control security levels
ASVS	4.2.1 – Ensure authorization is performed on all objects
CWE	CWE-425 – Direct request (forced browsing)

Mitigations	
	• Always perform authorization checks at the object level or path level so that you can be certain they cannot access resources they shouldn't

Summary

You've now covered the threat types described on the cards from the Elevation of Privilege suit in the Elevation of Privilege card deck. This, for obvious reasons, is also the trump suit for the game, not only for its name but also for the gravity of the risks these threats pose. These threats detailed flaws relating to supply chain attacks, insecure cloud solutions, containers and virtualization, poor documentation, XSS, and different forms of injection.

You should now understand the following:

- The need to have a vendor selection process and look at their track history
- The need to use MFA
- When to choose dedicated hardware
- The need to document your APIs thoroughly
- The need to verify assumptions you are making when using third-party APIs
- The need to validate and sanitize input and output as well as to encode output

In the next chapter, we will cover the privacy extension that turns **STRIDE** into **STRIPED** and includes privacy threats.

8
Privacy

Data privacy is about having sovereignty over your personal data and how that data is used. The **Organization for Economic Co-Operation and Development** (OECD) in 2013 defined eight principles of data privacy on which the **General Data Protection Regulation** (GDPR) was later based in 2016. Although GDPR only has seven principles (listed as follows), its documentation is noticeably similar to the OECD document:

- Lawfulness, fairness, and transparency
- Purpose limitation
- Data minimization
- Accuracy
- Storage limitation
- Integrity and confidentiality (security)
- Accountability

There were some important additions to GDPR, such as the right to be forgotten and that the personal data of European citizens should not be transferred to a country where there is weaker protection of it.

Figure 8.1: Personal information being shared without your consent

In this chapter, we will cover the threats described on the cards from the Privacy suit, which is an extension to the Elevation of Privilege card by Mark Vinkovits from LogMeIn.

When playing the cards from this suit, you should keep these principles in mind, as well as your privacy policy and your terms and conditions.

In this chapter, we're going to look at the regulations that should be observed to protect the privacy and the personal data of natural persons (i.e., you and I). This is a complex topic and is very often neglected and misunderstood.

By the end of the chapter, you should have a better understanding of what a customer should be informed of, what you should collect, how it can be used, and what rights the customer has regarding the treatment of data and its removal/destruction.

Before we dive into the cards, let's have a quick overview of the references that I will use in this chapter.

References used in the chapter

In this chapter, the references used will be to three documents that have had a strong influence in shaping data privacy worldwide:

- **GDPR**: `https://gdpr-info.eu/`

- **California Consumer Privacy Act (CCPA)**: `https://leginfo.legislature.ca.gov/faces/codes_displayText.xhtml?division=3.&part=4.&lawCode=CIV&title=1.81.5`

- **OECD privacy guidelines**: `https://legalinstruments.oecd.org/en/instruments/OECD-LEGAL-0188`

In addition, I have referenced the **Health Insurance Portability and Accountability Act (HIPAA)** (`https://aspe.hhs.gov/reports/health-insurance-portability-accountability-act-1996`) for the 9 of Privacy, which covers medical personal identifiable information. For the Ace of Privacy, I have included references to two official documents containing the line being taken on inference by the EU (`https://www.europarl.europa.eu/RegData/etudes/STUD/2020/641530/EPRS_STU(2020)641530_EN.pdf`) and the California Attorney General (`https://oag.ca.gov/system/files/opinions/pdfs/20-303.pdf`).

2 of Privacy

Your system does not ship by default with optimized, privacy-friendly settings.

Threat	
	When asking customers to accept your privacy policy, instead of using checkboxes requiring them to opt in and consent to their data being used for marketing purposes, you ask them to opt out.
GDPR	Chapter 1, Art. 4 – (11) Chapter 2, Art. 7
CCPA and HIIPA	*1798.140. Definitions – (h)*
OECD	*Part 2, 7. Collection Limitation Principle*

Mitigations	
	• If using checkboxes to ask for consent, always set the default to be that consent hasn't been given; that way, the customer actively chooses to participate

3 of Privacy

Your system is not able to properly handle the withdrawal of consent or objection to processing.

Threat	
	You're giving subjects a means to consent to you collecting their data, without giving them some way to back out or a way that is unnecessarily difficult.
GDPR	Chapter 2, Art. 7
CCPA and HIIPA	*1798.120. Consumers' Rights to Opt Out of Sale or Sharing of Personal Information* *1798.121. Consumers' Right to Limit Use and Disclosure of Sensitive Personal Information*
OECD	N/A

Mitigations	
	• A subject should be able to withdraw their consent; ensure this is possible
	• Ensure your system adheres to any regulatory requirements on the retention of a subject's data, even after a customer has withdrawn their consent
	• Respond to the subject, informing them of any actions taken

4. of Privacy

Your system collects consent but does not document aspects as to how, when, and what consent was provided.

Threat	
	You have a privacy policy or some terms you ask subjects to accept, but you do not version these documents, so if the policy changes, the conditions that were accepted are no longer in the policy. In addition, you also don't know what the conditions were when the subjects signed up.
GDPR	Chapter 2, Art. 5 – 1. (a)
	Chapter 2, Art. 5 – 1. (b)
CCPA and HIIPA	*1798.100. General Duties of Businesses that Collect Personal Information*
OECD	*Part 2, 9. Purpose Specification Principle*
	Part 2, 12. Openness Principle

Mitigations	
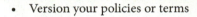	• Version your policies or terms • Store the version of the policy or terms against the subject's identifier so that you know what they agreed to • Store the date on which they agreed to the policy or terms • Store information on how they gave their consent (e.g., a web form, email, or snail-mail)

5. of Privacy

Personal data in your system is missing pointers to data subjects; hence, the data is forgotten when the owner is deleted or makes an access request.

Threat	
	You're collecting data, but that data isn't connected to the subject's user or ID, so when a subject is removed from the system, some of their data can become orphaned. It may also be impossible to retrieve a subject's data to show them exactly what data you have stored on them if they ask.
GDPR	Chapter 2, Art. 5 – 1. (e) Chapter 3, Art. 15 Chapter 3, Art. 16 Chapter 3, Art. 17
CCPA and HIIPA	1798.105. Consumers' Right to Delete Personal Information 1798.110. Consumers' Right to Know What Personal Information Is Being Collected. Right to Access Personal Information 1798.130. Notice, Disclosure, Correction, and Deletion Requirements
OECD	Part 2, 13. Individual Participation Principle

Mitigations	
	• There should be referential integrity constraints or equivalent that force the removal of child data before removing the parent

6. of Privacy

Your system collects more personal data than is strictly necessary to fulfill the intended purpose.

Threat	
	You might be storing information about the age, date of birth, or gender of your customers, but the only use you have for this data is to profile them, for which you haven't asked their permission.
GDPR	Chapter 2, Art. 5 – 1. (c)
CCPA and HIIPA	1798.100. General Duties of Businesses That Collect Personal Information (a) (1) and (2)
OECD	Part 2, 8. Data Quality Principle

Mitigations	
	• Only collect data for purposes your customers have agreed to

7. of Privacy

Your system is not following through on personal data deletion in integrated third parties.

Threat	
	You have outsourced some of your data processing to a partner company, perhaps for accounting or order fulfillment reasons, but if a customer asks to be removed from your systems, you do not ask the partner company to do the same.
GDPR	Part 3, Art. 17–2 Part 3, Art. 19
CCPA and HIIPA	1798.105. Consumers' Right to Delete Personal Information (c) (1)
OECD	N/A

Mitigations	
	• Your partner companies must be required to remove customer data that you ask them to • You must notify your partner companies if a customer asks for their data to be removed • Stipulate third-party obligations as part of vendor agreements, including any penalties for not complying with these obligations

8. of Privacy

Your system collects personal data without being able to name the specified, explicit, and legitimate purpose it is used for.

Threat	
	You have many applications that all use the same database and perform different tasks, but you do not track which elements of the collected data they use. So, if a subject asks how you process their data, you're not able to tell them with any certainty.
GDPR	Part 2, Art. 5–1. (b) Part 2, Art. 5–1. (c)
CCPA and HIIPA	1798.100. General Duties of Businesses that Collect Personal Information (c)
OECD	Part 2, 8. Data Quality Principle

Mitigations	
	• Document which data elements each of your applications processes • Audit which data elements each of your applications processes • Declare in the policies and terms what the data is collected for and how it is processes • Version your policies and terms • Review and update this information regularly • Ask for approval of changes from a subject before changing the way you process their data • Store the version of the policies and terms that the subject approved

9. of Privacy

Your product team avoids required controls for personal data as they move it outside of regulated and hardened environments.

Threat	
	If you work with PII, health, or financial data, you will have to meet certain regulatory requirements, such as encrypting personal information at rest and in transit. Members of your product team copy the production database onto their laptops to test new features without encrypting it locally.
GDPR	N/A
CCPA and HIIPA	https://aspe.hhs.gov/reports/health-insurance-portability-accountability-act-1996
OECD	N/A

Mitigations	
	• Production data should not be visible to the product team as part of their day-to-day operations • Production data should not be used for testing • Data should always be processed, with security measures approved for its classification or stronger.

10. of Privacy

Your system does not implement the erasure or anonymization of personal data once the legal ground for processing has been withdrawn.

Threat	
	When a subject withdraws consent for their personal data to be processed, you don't have a process in place to do this and aren't aware of what data must not be removed because of legal requirements on its retention that conflict with the rights of the subject and what you must remove because there is no legal requirement to retain it and it therefore must be removed.
GDPR	Chapter 3, Art. 17
CCPA and HIIPA	1798.105. Consumers' Right to Delete Personal Information
OECD	N/A

Mitigations	
	• Document which regulations affect which elements of the data you collect
	• Create a process to archive data that must be retained, which should include its safe disposal once the regulatory requirement to retain it expires
	• Define a process for the removal or anonymization of any remaining data
	• Automate and test the process of removing a subject's data

Jack of Privacy

Your system processes personal data in countries or with third parties that have weak privacy standards.

Threat	
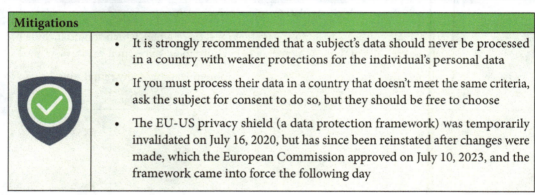	You process data of European citizens in a country not in the following list (Andorra, Argentina, Canada (only commercial organizations), Faroe Islands, Guernsey, Israel, Isle of Man, Jersey, New Zealand, Switzerland, Uruguay, Japan, the United Kingdom, and South Korea), without the explicit consent of the citizens or some other exception that permits its processing in a country where the protection of personal data cannot be adequately assured.
GDPR	Chapter 5, Art. 45
CCPA and HIIPA	1798.115. Consumers' Right to Know What Personal Information Is Sold or Shared and to Whom 1798.120. Consumers' Right to Opt Out of Sale or Sharing of Personal Information
OECD	*Part 4, Basic Principles of International Application: Free Flow and Legitimate Restrictions*

Mitigations	
	• It is strongly recommended that a subject's data should never be processed in a country with weaker protections for the individual's personal data • If you must process their data in a country that doesn't meet the same criteria, ask the subject for consent to do so, but they should be free to choose • The EU-US privacy shield (a data protection framework) was temporarily invalidated on July 16, 2020, but has since been reinstated after changes were made, which the European Commission approved on July 10, 2023, and the framework came into force the following day

Queen of Privacy

Your system processes personal data in a way that is not described in the privacy notice.

Threat	
	You perform profiling of your customers based on their purchases; however, you only have consent to use their data for the purpose of fulfilling their orders.
GDPR	Chapter 2, Art. 5–1 (b)
	Chapter 2, Art. 5–1 (c)
CCPA and HIIPA	1798.121. Consumers' Right to Limit Use and Disclosure of Sensitive Personal Information
OECD	*Part 2, 9. Purpose Specification Principle*
	Part 2, 10. Use Limitation Principle
	Part 2, 13. Individual Participation Principle pt. (d)

Mitigations	
	<ul><li>If you intend to process a subject's data in some way, it must be explicitly declared in your privacy notice</li><li>Version your privacy notice</li><li>Store information about the version of your privacy notice that a subject approved</li><li>Update your privacy notice with every change in the usage of a subject's data</li><li>Ask the subject for their approval of changes to the privacy notice</li><li>Audit periodically how the subject's data is used to ensure that it is aligned with the conditions they accepted</li></ul>

King of Privacy

Your system reuses personal data collected for a specific purpose for another, non-compatible purpose.

Threat	
	You use your customer data for direct marketing and telesales, but your customers have only given consent to their data to be used for the fulfillment of orders.
GDPR	Chapter 2, Art. 5 – 1. (b)
CCPA and HIIPA	1798.100. General Duties of Businesses That Collect Personal Information (a) (1) and (2)
OECD	*Part 2, 10. Use Limitation Principle*

Mitigations	
	• Only use a subject's data for the purpose they have approved, and nothing more

Ace of Privacy

You've invented a new privacy violation.

Threat	
	You purchase datasets from third parties and then correlate this data with existing data about your customers to enrich your dataset. This allows you to infer other information about your customers that perhaps they would not want you to know.
GDPR	`https://www.europarl.europa.eu/RegData/etudes/STUD/2020/641530/EPRS_STU(2020)641530_EN.pdf` (*page 6, paragraph 6*)
CCPA and HIIPA	`https://oag.ca.gov/system/files/opinions/pdfs/20-303.pdf`
OECD	N/A

Mitigations	
	• You should ask a subject's permission before enriching their data from other sources and inferring information about them

Summary

You've now covered the threat types described on the cards from the Privacy suit in the Elevation of Privilege with Privacy card deck. These threats deal with the fundamental rights and liberties of individuals. Protecting the privacy and personal information of individuals is something we should feel ethically bound to perform and is something that, in many cases, we are legally bound to perform. However, many organizations neglect this, either because they are unaware or, in some cases, deliberately looking to gain something, as the rewards outweigh the penalties. I implore you to act responsibly and ethically in this regard.

You should now understand the privacy principles defined by the OECD and the GDPR. You should know how to apply them, what rights you have as an individual, and, therefore, what rights your customers also have. You should also know what the customer needs to be informed of when collecting data and what you can do with it.

In the next chapter, we will cover the first of the **TRIM** categories of threats, **Transfer**.

9
Transfer

Transport of personal data across geopolitical or contractual boundaries can be quite troublesome because there are many things you need to consider. As already stated in *Chapter 8*, many of the laws and regulations around privacy today are based on the eight principles of privacy proposed by the **Organization for Economic Co-operation and Development (OECD)**. However, they also suggested that there should be a free flow of data between nations for as long as the following conditions apply:

- The other country follows the same guidelines

- The data controller has put in place sufficient safeguards to ensure a sufficient level of protection to meet the guidelines

Figure 9.1: Data being transferred around the world

The **General Data Protection Regulation (GDPR)** and many other regulations took this a step further, however, and applied it to their citizens' data irrespective of where that data is. The GDPR states the following:

> *"Any transfer of personal data which are undergoing processing or are intended for processing after transfer to a third country or to an international organisation shall take place only if, subject to the other provisions of this Regulation, the conditions laid down in this Chapter are complied with by the controller and processor, including for onward transfers of personal data from the third country or an international organisation to another third country or to another international organisation. All provisions in this Chapter shall be applied in order to ensure that the level of protection of natural persons guaranteed by this Regulation is not undermined."*

The preceding quote has been cited from the *EU General Data Protection Regulation (GDPR): Regulation (EU) 2016/679 of the European Parliament and of the Council of 27 April 2016 on the protection of natural persons with regard to the processing of personal data and on the free movement of such data, and repealing Directive 95/46/EC (General Data Protection Regulation), OJ 2016 L 119/1.*

You should have familiarized yourself with your company's privacy policy and associated procedures before you embark on designing software, sharing data for processing, or storing data. Your privacy policy should include information sometimes in an annex that specifies any data processors you use and their location. It should also include information about the data you collect, the purpose for which the data is being collected, and where you intend to process that data.

These are important because if your software is going to be collecting additional data or processing it for a purpose not listed, then you will need to ask customers to consent and potentially opt out if they do not agree to the new terms.

Before sharing with third-party processors, you should also verify that they are in the list of data processors you have shared with customers. If they are not, they are presumably new vendors and should go through an approval process in which they are vetted. At this point, once again, ask your customers whether they consent to their data being processed by this third party. Contracts with this third-party processor should contain their obligations with regard to transfer across borders as well as sharing, in the event that they have subcontractors, and any penalties for non-compliance.

In this chapter, we're going to look at the privacy regulations that should be observed with regard to the geographic location where you transfer, store, and process data.

References used in the chapter

As with the chapter on privacy (*Chapter 8*), in this chapter and the other T.R.I.M. chapters, the references used will be to four documents that have had a strong influence in shaping data privacy worldwide:

- **GDPR**: `https://gdpr-info.eu/`
- **California Consumer Privacy Act** (**CCPA**): `https://leginfo.legislature.ca.gov /faces/codes_displayText.xhtml?division=3.&part=4.&lawCode= CIV&title=1.81.5`
- **California Privacy Rights Act** (**CPRA**): `https://thecpra.org/`
- **OECD privacy guidelines**: `https://legalinstruments.oecd.org/en/ instruments/OECD-LEGAL-0188`

2. of Transfer I

The application uses an API, which makes it our data processor, but we don't know whether this is reflected in our API contract.

Threat	
	You're using a third-party service to profile your customers and, although your privacy policy states that you perform profiling, you don't know whether your privacy policy has been updated to reflect the use of a third party.
GDPR	Chapter 2, Art 5. - (a)
CCPA &	CCPA 1798.100. General Duties of Businesses that Collect Personal Information (d)
CPRA	CPRA SEC. 3. Purpose and Intent. (A)(1)
OECD	Paragraph 15(a)(i)

Mitigations	
	• Before making changes to your data handling, check with your legal department whether the changes need to be communicated to your customers
	• Ensure that you have read and understood your company policies
	• Ensure that you are aware of your contractual obligations

2. of Transfer II

The application uses an API, which makes it our data processor, but we don't know whether this is reflected in our API contract.

Threat	
	In the code of your application, you are calling a third-party service for the company you use to handle your deliveries and sending them the name and address of your customer. However, you don't know whether your customer was made aware that someone else would see their personal data.
GDPR	Chapter 3, Art. 13 – 1. (e)
CCPA &	CCPA 1798.100. General Duties of Businesses that Collect Personal Information (d)
CPRA	CPRA SEC. 3. Purpose and Intent. (A)(1)
OECD	Paragraph 15(a)(i)

Mitigations	
	• In your privacy policy, state who your data processors are and what data they will be processing

3. of Transfer

We provide an API that ingests personal data, but we do not know whether we are a data processor or a data controller, and it's not defined in our contracts.

Threat	
	You've exposed an API that processes personal data, but you aren't certain where this data comes from: is it from the subject, or is it from someone else who has collected the data from the subject? Who are the consumers of your API and do your contracts cater to those different types of consumers?
GDPR	Chapter 1, Art 4. – (7) and (8) Chapter 4, Art 24. Chapter 4, Art 28. Chapter 4, Art 29.
CCPA & CPRA	CCPA 1798.140. Definitions (ag)
OECD	N/A

Mitigations	
	• You must know and understand who the consumers of your service are to be able to better understand your role • You also need to document, in contracts with your customers, what your role is in handling the data

4. of Transfer

We call an API with personal data, but we do not know where the API is being hosted geographically.

Threat	
	An API you are calling is hosted in the cloud, but you have no idea what data center it's hosted in. As you are processing EU citizens' personal data, EU regulations apply. If the data is being processed in a country with less stringent protection of personal data, this would be in violation of GDPR.
GDPR	Chapter 5, Art. 44 Chapter 5, Art. 45 Chapter 5, Art. 46
CCPA & CPRA	CPRA SEC. 4. Section 1798.100 General Duties of Businesses that Collect Personal Information (d)(2)
OECD	Part 4, Basic Principles of International Application: Free Flow and Legitimate Restrictions 16 and 17

Mitigations	
	• Always ensure that you know where data is being processed • Be informed on what the data privacy laws are both for your country and the countries you work with

5. of Transfer

We export a database dump by writing a CSV file on an FTP site. What happens to the file after it has been downloaded is not our problem.

Threat	
	You're allowing third parties to download a dump of your database and have no record of who downloaded the data and no contract in place stipulating the third parties' obligations regarding the treatment of the data. The data is your responsibility for as long as you are the owner or controller. If you grant someone access to take a copy of that data, it is still your responsibility, and if a subject then asks for their data to be removed, you must approach anyone you have given a copy to and ask them to remove the data as well.
GDPR	Chapter 4, Art. 24
CCPA &	CCPA 1798.100. General Duties of Businesses that Collect Personal Information (d)
CPRA	CPRA SEC. 3. Purpose and Intent. (A)(1)
OECD	Part 2, Accountability Principal 14

Mitigations	
	• Always keep track of who has a copy of the data of which you are the custodian or owner • Have a process in place to ask for changes to the data they have • Prohibit contractually the sharing of any personal data for which you are responsible

6. of Transfer

Some of our systems are hosted outside the EU, but the service provider says that they take security very seriously, so that's fine.

Threat	
	You are hosting your data in a country with a military regime and the government may ask to see the data and the organization hosting the data may be obligated to grant access. The organizational security isn't sufficient. The nation must respect human rights and basic freedoms, have a supervisory body, and have international agreements in place for the hosting to be outside the EU.
GDPR	Chapter 5, Art. 44 Chapter 5, Art. 45 Chapter 5, Art. 46
CCPA & CPRA	CPRA SEC. 4. Section 1798.100 General Duties of Businesses that Collect Personal Information (d)(2)
OECD	Part 4, Basic Principles of International Application: Free Flow and Legitimate Restrictions 16 and 17

Mitigations	
	• Ensure that data is hosted in countries with privacy laws that are of the same strength or stronger than those in the GDPR

7. of Transfer

Our systems are being administered from outside the EU, but admin access is not personal data access, right?

Threat	
	You have not implemented the principle of least privilege on your systems. Due to that, an administrator has access to everything, whenever they want. So, by accessing this information, they are effectively exporting the information to another state outside of the EU.
GDPR	Chapter 5, Art. 44 Chapter 5, Art. 45 Chapter 5, Art. 46
CCPA & CPRA	CPRA SEC. 4. Section 1798.100 General Duties of Businesses that Collect Personal Information (d)(2)
OECD	Part 4, Basic Principles of International Application: Free Flow and Legitimate Restrictions 16 and 17

Mitigations	
	• Implement the principle of least privilege: you can only access what you need to access • Avoid allowing administrators to escalate their own privileges; another administrator should be required to grant them additional privileges • Any additional privileges should be granted only for the time necessary to perform the required task

8. of Transfer I

We send personal data over email, but only within the company, so that should be fine, right?

Threat	
	You're not encrypting connections to the mail server, so someone could listen in. You are also storing unencrypted emails in mailboxes on your exchange that may be read by an administrator.
GDPR	Chapter 4, Art. 32 - 1. (a)
	Chapter 4, Art. 32 - 2.
CCPA & CPRA	CCPA 1798.100. General Duties of Businesses that Collect Personal Information (e)
OECD	Part 2, 11. Security Safeguards Principle

Mitigations	
	• Require TLS to connect to the mail server
	• Require end-to-end encryption to ensure that only you and the recipient are able to read the content

8. of Transfer II

We send personal data over email, but only within the company, so that should be fine, right?

Threat	
	You haven't restricted access to the mailboxes on the server and administrators can read the content of communications.
GDPR	Chapter 4, Art. 32 - 4.
CCPA & CPRA	CCPA 1798.100. General Duties of Businesses that Collect Personal Information (e)
OECD	Part 2, 11. Security Safeguards Principle

Mitigations	
	• Restrict access to mailboxes on the server • Require approvals for access to be granted • Encrypt content when on the server

8. of Transfer III

We send personal data over email, but only within the company, so that should be fine, right?

Threat	
	The personal information you are sharing in emails can be forwarded and you no longer have control over who will see the PII.
GDPR	Chapter 2, Art. 5 – 1. (f) Chapter 4, Art. 32
CCPA & CPRA	CCPA 1798.100. General Duties of Businesses that Collect Personal Information (e)
OECD	Part 2, 11. Security Safeguards Principle

Mitigations	
	• Implement **Data Leak Prevention** (DLP) at the perimeter/outbound mail server to block content that shouldn't be shared from being shared

9. of Transfer

We provide an API to access personal data, and we do not control who can access this API.

Threat	
	Your API doesn't require any authentication, thereby making it public and open to use by anyone, so people can make requests without you knowing who they are or whether they should even be accessing your data.
GDPR	Chapter 2, Art. 5 – 1. (f) Chapter 4, Art. 32
CCPA & CPRA	CCPA 1798.100. General Duties of Businesses that Collect Personal Information (e)
OECD	Part 2, 11. Security Safeguards Principle

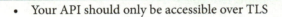

Mitigations	
	• Your API should only be accessible over TLS
	• Your API should require authentication
	• Credentials should be associated with a user or organization
	• Access should be logged
	• Your API should also require authorization so that you can specify what a consumer can access

Ace of Transfer

You have identified a new personal data flow out from your system.

Threat	
	Your organization is based in the US, but you offer goods and services to individuals located in the EU without having met all the GDPR requirements for the processing of their data.
GDPR	Chapter 1, Art. 3 – 2.
CCPA & CPRA	N/A
OECD	N/A

Mitigations	
	• Ensure that you are meeting GDPR requirements if you wish to continue trading with EU residents
	• Implement the EU-US Data Privacy Framework approved on the 10th of July 2023, which supersedes the EU-US Privacy Shield

Summary

You've now covered the threat types described on the cards from the Transfer suit of the Elevation of Privacy extension card deck. These threats deal with the fundamental rights and liberties of individuals with a focus on international flows of data. This is an important topic because the same levels of protection are not necessarily enforced in other countries, to the extent that not respecting these regulations could put someone's life at risk if, for example, they are a refugee.

You should now have an understanding of how these privacy regulations affect your organization with regard to the collection, processing, and storage of data.

In the next chapter, we will cover the Retention/Removal category of threats from T.R.I.M. and the Elevation of Privacy extension for the game.

10
Retention/Removal

Many of the data privacy regulations require that data should be retained or removed based on certain criteria:

- Legal requirements concerning financial data may stipulate the need for it to be retained for up to 10 years

- The storage limitation principle implies that **Personally Identifiable Information** (**PII**) should only be retained for the time necessary to process the data for the purpose intended or for as long as the data is in the public interest, or for research but after it has been anonymized

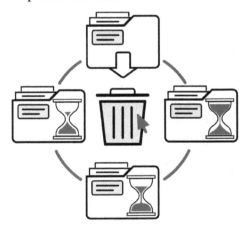

Figure 10.1: Files are retained until no longer needed and are then destroyed

In this chapter, we're going to look at the privacy regulations that should be observed with regard to retention and the removal of data from your systems.

As with the chapter on privacy, in this chapter and the other T.R.I.M. chapters, the references used will be to the GDPR, CCPA, CPRA, and OECD documents that have had a strong influence in shaping data privacy worldwide.

We will cover under what circumstances you should be retaining a subject's personal data, for how long you should retain it, what circumstances give you the right to retain a subject's personal data, and under what circumstances you are obliged to remove a subject's personal data. By the end of the chapter, you should understand both your rights and those of the subject.

2. of Retention/Removal

Users' file uploads containing personal data are saved to temporary files on the frontend.

Threat	
	You've implemented hot-desking in your office since moving to a hybrid working model, and when your staff uploads personal data to the HR system, that personal data is stored on the local system temporarily but never cleaned up. This has made it possible for their colleague to read the files they uploaded because this computer is now shared.
GDPR	Chapter 4, Art. 32 – 1 (b) Chapter 1, Art. 4 – (12) Chapter 2, Art. 5 – 1 (f)
CCPA & CPRA	CCPA 1798.100. General Duties of Businesses that Collect Personal Information (e)
OECD	Part 2, 11. Security Safeguards Principle

Mitigations	
	• Temporary files should be removed when they are no longer needed because the task is complete • Data stored on workstations should not be accessible by another user of the same workstation

3. of Retention/Removal I

All personal data goes into a large pile in the cloud, and going through it to find individual records would cost a fortune in retrieval and outbound data transfer fees.

Threat	
	You haven't created an index of your data, and it is stored as flat files, so finding information requires downloading the files and then searching locally.
GDPR	Chapter 2, Art. 5 – 1. (e) Chapter 4, Art. 32 – 1. (b) Chapter 4, Art. 32 – 1. (c)
CCPA & CPRA	N/A
OECD	N/A

Mitigations	
✅	• Index/catalog your data to facilitate fast and efficient searching. Similar to having a table of contents or the index at the back of a book, having an index containing keywords or similar linked to filenames can help you find the records you need.

3. of Retention/Removal II

All personal data goes into a large pile in the cloud, and going through it to find individual records would cost a fortune in retrieval and outbound data transfer fees.

Threat	
	To reduce costs, you decided to use a long-term storage solution that is inexpensive for rarely accessed data but very costly for frequent reads and writes. As you deal with large amounts of personal data, you are making lots of writes and are frequently required to perform searches that involve lots of reads.
GDPR	Chapter 4, Art. 32 – 1. (b)
	Chapter 4, Art. 32 – 1. (c)
CCPA & CPRA	N/A
OECD	N/A

Mitigations	
	• Review your requirements and the conditions for each of the cloud provider's storage solutions to determine the most adaptable solution for your needs
	• Index your data to facilitate fast and efficient searching

4. of Retention/Removal

We store personal data on disk, even though we only need it temporarily and could just cache it in memory.

Threat	
	You're storing data on a disk that you don't need, which is costing you money, putting you at risk, and can only be used for the purpose it was originally intended according to data privacy regulations. This is temporary data and could be queued until it is processed and then destroyed.
GDPR	Chapter 2, Art. 5 – 1. (e)
CCPA & CPRA	CCPA 1798.100. General Duties of Businesses that Collect Personal Information (a)(3)
OECD	N/A

Mitigations	
	• Familiarize yourself with the data privacy regulations • Avoid using long-term storage for short-term usage • Don't retain it longer than necessary • Don't put yourself at risk by unduly increasing your attack surface • Use eviction policies to ensure that data is removed automatically after an agreed period if it must be stored

5. of Retention/Removal

When changing data, we retain all old data in order to be able to show what has been changed.

Threat	
	You are retaining data and never deleting it; instead, you are flagging it as deleted so that you don't lose anything, which is wasteful of resources and goes against your declared retention policy.
GDPR	Chapter 2, Art. 5 – 1. (d)
CCPA & CPRA	CCPA 1798.106. Consumers' Right to Correct Inaccurate Personal Information (c)
OECD	N/A

Mitigations	
	• Develop a backup policy. If changes are made accidentally or maliciously, the data can be restored from backup • Eliminate backups after the agreed backup retention period has expired

6. of Retention/Removal

The personal data is stored on a blockchain. We can't delete it at all.

Threat	
	You are storing data in a blockchain. If a subject asks for their data to be deleted, you aren't able to fulfill their request, which is a violation of privacy regulations in some regions.
GDPR	Chapter 3, Art. 17 – 1. (b) Chapter 3, Art. 17 – 3.
CCPA & CPRA	CCPA 1798.105. Consumers' Right to Delete Personal Information (c)(1)
OECD	N/A

Mitigations	
	• Choose very carefully what you put in a blockchain as, for the entire life of the blockchain, the information cannot be removed • Alternatively, use smart contracts in which one of the events deletes the content or revokes all access, a bit like crypt-shredding

7. of Retention/Removal

Consent is a checkbox, but to withdraw the consent and remove your data, you need to email us.

Threat	
	You require your customers to send an email to withdraw their consent, but when you ask them to consent, they only need to click a checkbox. *Article 7* of GDPR *Conditions for Consent* condition number 3 states that "*It shall be as easy to withdraw consent as to give consent*," so by requiring an email be sent, the process is more complex than it should be for the subject. Other regulations also suggest a similar approach.
GDPR	Chapter 2, Art 7. – 3.
CCPA & CPRA	CCPA 1798.135. Methods of Limiting Sale, Sharing, and Use of Personal Information and Use of Sensitive Personal Information (b)(2)(A)
OECD	N/A

Mitigations	
	• If you have a checkbox to grant consent, you must have a method of similar simplicity to revoke consent

8. of Retention/Removal

We have not defined a specific retention time for personal data, but we can delete it if someone asks us to.

Threat	
	You're not declaring how long you intend to retain the data and you're retaining data beyond the time required to process the data for the purpose it was intended or archiving it beyond the time it is of public interest.
GDPR	Chapter 3, Art 13. – 2. (a)
CCPA & CPRA	CCPA 1798.100. General Duties of Businesses that Collect Personal Information (a)(3)
OECD	N/A

Mitigations	
	• Define a data retention policy in which you specify the retention period • If you don't define a data retention policy, retain the data for as long as the subject is a customer, or as long as it takes to process the data unless there is some legal impediment requiring its retention

9. of Retention/Removal

Yes, we have defined a retention time for personal data; it's defined by the IT department based on disk space usage.

Threat	
	You are only removing data as a means to save disk space, which does not allow you to define the retention period, and neither does it meet the requirements of the different regulations. GDPR and other regulations state that *"data should be kept in a form which permits identification of data subject no longer than is necessary for the purposes for which the personal data are processed"* – unless it's being archived because the data is in the public interest or for scientific research.
GDPR	Chapter 2, Art. 5 – 1. (e) Chapter 3, Art 13. – 2. (a)
CCPA & CPRA	CCPA 1798.100. General Duties of Businesses that Collect Personal Information (a)(3) CCPA 1798.100. General Duties of Businesses that Collect Personal Information (c) CCPA 1798.125. Consumers' Right of No Retaliation Following Opt-Out or Exercise of Other Rights (b) (1)
OECD	N/A

Mitigations	
	• Remove the data when no longer required for the purpose it was intended • Specify a maximum retention period – for example, the data will be kept for a maximum of 2 years from the date it was last used – in the event that the subject requires access to their historical data

10. of Retention/Removal

We cannot remove personal data, as the database schema requires the data to be there.

Threat	
	You're linking personal data to orders in your system and, because of referential integrity constraints, removing a customer would also remove their orders and the products they ordered from the database.
GDPR	Chapter 3, Art. 17 – 1. (b)
CCPA & CPRA	CCPA 1798.105. Consumers' Right to Delete Personal Information (c) (1)
	CCPA 1798.125. Consumers' Right of No Retaliation Following Opt-Out or Exercise of Other Rights (b) (1)
OECD	N/A

Mitigations	
	• Ensure that the coupling between data subjects and your organizational data is such that you can eliminate customer records without damaging your historical and operational data
	• Include customer IDs without using them as a foreign key

Jack of Retention/Removal

We have defined a retention time for personal data, but that's only a policy. There is no technical system that enforces it.

Threat	
	You're telling data subjects that their data will be removed after a period of 2 years, but it is not removed. If you suffered a data breach, data that should have been removed 6 years ago would also be leaked.
GDPR	Chapter 3, Art 13. – 2. (a) Chapter 4, Art. 25 – 2.
CCPA & CPRA	CCPA 1798.100. General Duties of Businesses that Collect Personal Information (a) (3) CCPA 1798.100. General Duties of Businesses that Collect Personal Information (c)
OECD	N/A

Mitigations	
	• If you are declaring something in a policy, always make sure you are following through on it • Ignorance that the data was retained beyond the period stated in your policy is not a mitigation • If not automated, put alerting in place so that a process of data cleansing can be performed • Where possible, automate your data cleansing but ensure the process is thoroughly tested before putting it into production

Ace of Retention/Removal

You have found a new personal data storage location that you did not know existed.

Threat	
	It costs less to retain customer personal data than to delete it, so you have been gradually accumulating large amounts of customer personal data.
GDPR	Chapter 3, Art. 17 – 1. (b)
CCPA & CPRA	CCPA 1798.105. Consumers' Right to Delete Personal Information (c) (1) CCPA 1798.125. Consumers' Right of No Retaliation Following Opt-Out or Exercise of Other Rights (b) (1)
OECD	N/A

Mitigations	
	• Where possible, remove data when no longer required for processing • If it is more economically viable to anonymize the data, then this may be an alternative

Summary

Most privacy regulations require that you only keep data as long as it's needed. Keeping data that you cannot use or are not using may be costing you money but certainly leaves you open to unnecessary risk. The privacy regulations also usually specify that the subject has the right to have their data removed on request.

Therefore, when designing software, it makes sense to consider building in a mechanism for removing data when no longer required or at the request of the subject. Anonymization is another option, but it must be done correctly so that the data cannot be linked back to the subject, even when correlating multiple datasets.

You've now covered the threat types described on the cards from the Retention/Removal suit of the Elevation of Privacy extension card deck. These threats deal with the storage limitation principle and a subject's right to be forgotten. Unless it is in the public interest, for scientific research, freedom of expression, or because of a legal obligation, you should remove a subject's data once it has served its purpose. You should also remove the subject's data upon request.

You should now understand how these privacy regulations affect your organization with regard to the retention and removal of a subject's personal data.

In the next chapter, we will cover the Inference category of threats from T.R.I.M. and the Elevation of Privacy extension for the game.

11
Inference

The result of inference of personal data from other personal data, for example, through correlation, is still considered personal data. Knowing that two people frequently visit multiple locations at the same time and often leave at the same time can be an indicator that the two people know one another. Furthermore, if they often arrive and leave at the same time it could even be inferred that they were meeting at these locations.

Figure 11.1: From incomplete data you can infer the result

In this chapter, we're going to look at the privacy regulations that should be observed with regard to inference by your systems.

As with the chapter on privacy, in this chapter and the other T.R.I.M. chapters, the references used will be to the GDPR, CCPA, CPRA, and OECD documents that have had a strong influence on shaping data privacy worldwide. In addition, this following document explains the relationship between the GDPR and artificial intelligence: *The impact of the General Data Protection Regulation (GDPR) on artificial intelligence – EPRS European Parliament Research Service*, available at https://www.europarl.europa.eu/RegData/etudes/STUD/2020/641530/EPRS_STU(2020)641530_EN.pdf.

We will cover inference and how the different privacy regulations are impacted by it, the use of profiling, and the difference between automated and manual decision making based on inferred data. By the end of the chapter, you should understand where to draw the line.

2. of Inference

You use a common identifier across all systems and also expose this to third parties.

Threat	
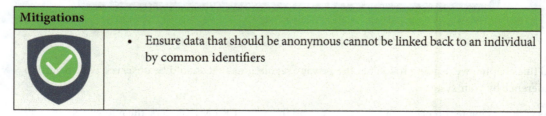	You are collecting medical data for clinical trials, as well as location data from satellite vehicle-tracking anti-theft devices. One of your customers in the insurance industry has requested access to both datasets. They are able to correlate the data because the identifiers are evidentially related and therefore infer that an individual is doing clinical trials, which permits their insurance firm to increase their premiums for health insurance.
GDPR	Chapter 2, Art. 9 – 1. Chapter 2, Art. 9 – 2. Chapter 3, Art. 22 – 1. Recital 71
CCPA & CPRA	CCPA 1798.140. Definitions (m)(1) CCPA 1798.140. Definitions (m)(2) CCPA 1798.140. Definitions (m)(3)
OECD	N/A

Mitigations	
	• Ensure data that should be anonymous cannot be linked back to an individual by common identifiers

3. of Inference

Your geolocation data is as accurate as possible, even if you only really need to know which city the user is from.

Threat	
	If you are collecting the most detailed geo-location data, you should have a good reason for doing so, you should have communicated that reason to the subject, you should only collect it when it is necessary to collect it, and you should likewise only retain it for as long as is strictly necessary. If recorded continually and stolen, that data could be used to determine the daily routine of a subject, from which someone could infer any number of things including their beliefs if they visit a place of worship, their political affiliations, their medical situation, and so on.
GDPR	Chapter 1, Art. 4 – 4. Chapter 2, Art. 5 – 1. (b) Chapter 2, Art. 9 – 1.
CCPA & CPRA	CCPA 1798.185. Regulations (a)(13) CCPA 1798.140. Definitions (w) CCPA 1798.140. Definitions (ae)(1)(C)
OECD	N/A

Mitigations	
	• Only collect the level of detail required for your specific need • Only collect what you have said you will collect • Don't keep hold of data you don't need; it will waste resources and increase risk

4. of Inference

You use your users' names or email addresses as reference keys between systems, even though we could use random identifiers.

Threat	
	You are using names or email addresses to reference your data, which means that the data is not easily anonymized. You may end up with name clashes that result in incorrect data. Email addresses may be reused if an email account is abandoned or a shared account is used, again resulting in incorrect data.
GDPR	Chapter 2, Art. 5 – 1. (d) Chapter 2, Art. 5 – 1. (e)
CCPA & CPRA	CCPA 1798.140. Definitions (m)(1) CCPA 1798.140. Definitions (m)(2)
OECD	N/A

Mitigations	
	• Using random identifiers can facilitate anonymization • It could also avoid potential identity clashes if email addresses were recycled or shared

5. of Inference

You use national ID numbers or **Social Security Numbers** (**SSNs**) as identifiers because they are conveniently unique.

Threat	
	You are using national ID numbers or SSNs, which are very sensitive data, to identify records. Using them as identifiers means they must be stored in clear text. This sort of data should always be encrypted at rest as well as in transit.
GDPR	Chapter 2, Art. 5 – 1. (f) Chapter 2, Art. 5 – 2. Chapter 4, Art. 32 – 1.
CCPA & CPRA	CCPA 1798.100. General Duties of Businesses that Collect Personal Information (e)
OECD	N/A

Mitigations	
	• National ID numbers and SSNs should not be used other than for the purpose they were intended • They should be encrypted both at rest and in transit

6. of Inference

You use identifiers in our web links. These identifiers are leaked in browsers' referrer headers and get logged by redirectors and URL shorteners.

Threat	
	Your Java web application uses a JSESSIONID that it includes in the URL. Before redirecting to other sites, be careful what you include in the URL. Most modern browsers will only send the origin to other sites.
GDPR	Chapter 1, Art. 4 – 12. Chapter 2, Art. 5 – 1. (f) Chapter 2, Art. 5 – 2.
CCPA & CPRA	CCPA 1798.100. General Duties of Businesses that Collect Personal Information (e)
OECD	N/A

Mitigations	
	• To be on the safe side, it's a good practice to include the following header to force the browser to only send the origin to other sites: `Referrer-Policy: strict-origin-when-cross-origin` • Using POST requests instead of GET requests would be the preferred method so that data isn't exposed in the URL

7. of Inference

There is no review process for introducing new trackers or advertising providers onto your web pages; whatever your designers like, or marketing sells, will be used.

Threat	
	You don't have a review process for new trackers or ad providers, so you are leaving yourself open to supply chain attacks and data leakage. A provider may be harvesting data about your customers without them ever having clicked on a link.
GDPR	Chapter 3, Art. 21 – 2. Chapter 3, Art. 21 – 3. Chapter 4, Art. 32 – 2. Chapter 4, Art. 32 – 4.
CCPA & CPRA	CCPA 1798.100. General Duties of Businesses that Collect Personal Information (d)(2) CCPA 1798.140. Definitions (h) – Dark Patterns CCPA 1798.185. Regulations (a)(20)(C)(i) CCPA 1798.110. Consumers' Right to Know What Personal Information is Being Collected. Right to Access Personal Information CCPA 1798.115. Consumers' Right to Know What Personal Information is Sold or Shared and to Whom
OECD	N/A

Mitigations	
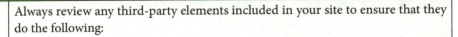	Always review any third-party elements included in your site to ensure that they do the following: • Respect your privacy policy • Don't contain malicious code • Ensure your vendor contracts contain penalties for misuse of data you are responsible for

8. of Inference

Your telemetry is tied to the users, even though your analytics couldn't care less who the user actually is.

Threat	
	You're storing data that is linked to your users and doesn't need to be. This puts you at risk in the event someone gains unauthorized access to the data.
GDPR	Chapter 2, Art. 5 – 1. (e)
CCPA & CPRA	CCPA 1798.100. General Duties of Businesses that Collect Personal Information (c)
OECD	N/A

Mitigations	
	• You should anonymize the data as the contextual information about a given user isn't of importance in your analysis

9. of Inference

A neural network makes customer-related decisions, but nobody can really explain to the customers what the model is based on.

Threat	
	If you are using a neural network to make decisions based on the data of EU citizens, the GDPR requires that you are able to explain why that neural network made a certain decision about them from their data. If you are unable to explain the model itself, it is impossible to explain why the system may have made such a decision.
GDPR	Chapter 1, Art. 4 – 4. Chapter 3, Art. 22 – 1. Recital 71
CCPA & CPRA	CCPA 1798.185. Regulations (a)(16)
OECD	Part 2, 7. Collection Limitation Principle

Mitigations	
	• When using neural networks, there must be a clear understanding of what the model is based on • You should record which features of the subject's data were used for training • Any hypotheses that were tested should also be documented • Any work performed to understand, prove, or disprove the findings of the model should also be documented

10. of Inference

You do not make any checks on personal data before you use it for training machine learning models.

Threat	
	You are not checking personal data before using it in training, so you cannot expect your model to be accurate. Therefore, some of the decisions it will make will also not be accurate, which implies it may unfairly make decisions about a subject, such as mortgage approvals/refusals. A requirement of many data privacy regulations is that the data be kept accurate. It is also possible that an attacker may modify the data deliberately to poison the model and you could be leaving yourself open to injection attacks.
GDPR	Chapter 2, Art. 5 – 1.(d) Chapter 2, Art. 5 – 1.(f)
CCPA & CPRA	N/A
OECD	Part 2, 8. Data Quality Principle

Mitigations	
	• Ensure that the data you are using is accurate and up to date • Ensure your models are trained and tested using different samples from a given dataset • Verify outliers

Ace of Inference

You have found a new place where you can replace personal data with a random identifier.

Threat	
	A customer has withdrawn consent for the processing of a subset of their data or for processing it in a certain way. You decided to keep the data and performed some anonymization on it, however, because of some shared characteristics/features in your datasets, it turns out it is only pseudonymized and it is possible to infer information the user chose not to reveal when withdrawing their consent.
GDPR	Chapter 2, Art. 7 – 3.
CCPA & CPRA	CCPA 1798.121. Consumers' Right to Limit Use and Disclosure of Sensitive Personal Information
OECD	N/A

Mitigations	
	• If you cannot entirely anonymize the data, remove it • Anonymizing the data ensures that it cannot be re-connected such that the combination of characteristics/features makes it possible to infer the identity of the subject

Summary

Inference is often used in profiling users to make decisions about products that may potentially interest them, their political persuasions to influence their voting, their sexual orientation, or any number of other very personal matters. By correlating multiple datasets, it becomes easier to build a complete picture of a person and it may even be possible to infer things about them that they don't yet know themselves.

When someone is undecided about which of two products to buy or which political party to vote for, inference then becomes a very powerful tool, but perhaps isn't entirely ethical. For example, this information can then be used to manipulate their decision with targeted messages based on prior knowledge of what can influence them. It is somewhat Machiavellian.

Inference may also be used in making automated decisions in the hiring process such as to filter CVs. In Europe, it is questionable whether this is legal because you could be unknowingly discriminating against certain categories of applicant without even being aware of it. The model is making decisions you may not be able to explain, even though you legally have an obligation to.

Inferences can be and often are incorrect, because the data being fed into the model is itself incorrect or there are other factors involved about which the model doesn't know. For example, there may be multiple family members using a single account to make purchases for online shopping. You may have a shared device in the living room and its search history could contain searches carried out by your daughter's best friend, your partner, or your neighbor, so making inferences about you based on this information is probably not going to give an accurate result.

So, be very careful when using inference to consider the ethical implications, the legal aspects, and the accuracy of the data being used to make the inferences. Ask yourself these simple questions:

- Would you want someone to infer the thing in question about you?

- Would you be happy if it was inferred about you incorrectly?

- Would this be an infringement of your privacy?

You've now covered the threat types described on the cards from the Inference suit of the Elevation of Privacy extension card deck. These threats deal with conclusions being drawn from data points relating to a subject based on statistical observations of historical data relating to other subjects. We have learned in this chapter that profiling and automated decision-making can be problematic from a privacy standpoint. We should consider there will be outliers (exceptions to the rule) and so mistakes will be made, and we should allow for this when dealing with data relating to individuals and be ready to handle appeals.

In the next chapter, we will cover the minimization category of threats from T.R.I.M. and the Elevation of Privacy extension for the game.

12

Minimization

Minimization is the act of ensuring that the data being gathered is only what is needed for the purpose intended and what the subject consented to, without gathering any unnecessary additional information.

Article 5 of the GDPR, *Principles relating to the processing of personal data*, states in Item 1 the following: "*Personal data shall be: (c) adequate, relevant and limited to what is necessary in relation to the purposes for which they are processed ('data minimization').*"

Figure 12.1: The fields we don't require are redacted

In this chapter, we're going to look at the privacy regulations that should be observed with regard to limiting data collected to only that necessary for the task at hand (minimization) by your systems.

As with the chapter on privacy in this chapter and the other TRIM chapters, the references used will be the GDPR, CCPA, CPRA, and OECD documents that have had a strong influence in shaping data privacy worldwide.

We will now cover threats related to minimization and the collection limitation principle. By the end of the chapter, you should understand what to collect, what not to collect, and why collecting irrelevant information increases risk without adding value.

2. of minimization

We put absolutely everything in the audit log so that we can positively audit all personal data activities.

Threat	
	Audit logs should contain enough information to identify who or what performed actions, when they did so, and what was impacted. You're logging more than is needed, which is wasteful of resources, goes against the principle of minimization, and leaves you more exposed.
GDPR	Chapter 2, Art. 5–1 (f) Chapter 2, Art. 5–2 Chapter 4, Art. 32–1
CCPA and CPRA	N/A
OECD	*Part 3, 15. A Data Controller Should (iv)* *Part 3, 15. A Data Controller Should (vi)*

Mitigations	
	• Only collect the data you need for the purpose you are collecting it • Run periodic company-wide checks to audit what is being collected and ensure the data you collect is necessary

3. of minimization

Our testing data is a month-old copy from production. Fake data just does not have the same feel to it.

Threat	
	You use real data for testing without anonymizing it, which could potentially give employees visibility to personal data they should otherwise not have access to. It also is open to potential theft. Often, testing environments aren't offered the same protection as production environments and are, therefore, more open to being breached and data being stolen.
GDPR	Chapter 2, Art. 5 – 1 (b) Chapter 2, Art. 5 – 1 (f) Chapter 2, Art. 6 – 1
CCPA and CPRA	*CCPA 1798.100. General Duties of Businesses that Collect Personal Information (a)(1)* *CCPA 1798.100. General Duties of Businesses that Collect Personal Information (a)(2)*
OECD	*Part 2, 9 Purpose Specification Principle* *Part 2, 10 Use Limitation Principle*

Mitigations	
	• Where possible, generate fake test data. You can use lookup dictionaries of values, such as a list of first and last names, and randomize the ones being chosen to facilitate this. There are also tools available that, when given a schema, will generate random test data for you. • If it would be over-complicated to generate fake data for your testing, anonymize a copy of real data, removing any identifying/identifiable information.

4. of minimization

Our website does not work at all with an ad blocker.

Threat	
	You decided that because ad blockers block tracking and analytics information, your site won't work if users have activated an ad blocker in their browser, which means you force them to choose whether they wish to maintain their privacy. You could stand losing customers by making this choice.
GDPR	N/A
CCPA and CPRA	CCPA 1798.125. Consumers' Right of No Retaliation Following Opt Out or Exercise of Other Rights
OECD	N/A

Mitigations	
	• Your site should allow users to use ad blockers if they want to maintain their privacy
	• If you don't want them to use ad blockers, be very careful to explain what data you are collecting, how it is used, whether you are tracking, and whether you are using any analytics

5. of minimization

We send personal data to an API, even though we believe it is not really being used for anything.

Threat	
	One of the third-party APIs you rely on includes fields that don't seem relevant or necessary for the API to elaborate the analysis you're asking, but you're including it because it is a required field.
GDPR	Chapter 2, Art. 5 – 1 (a) Chapter 2, Art. 5 – 1 (b) Chapter 2, Art. 5 – 1 (c)
CCPA and CPRA	CCPA 1798.100. General Duties of Businesses that Collect Personal Information (a)(1)
OECD	Part 2, 9. Purpose Specification Principle Part 2, 10. Use Limitation Principle

Mitigations	
	• Check with the third party why the data is required • If it is not required, ask them to change the API so that the field isn't required • Don't send the data • Send a dummy value for all requests

6. of minimization

We'll just block the EU and California from our site. We've got enough customers elsewhere.

Threat	
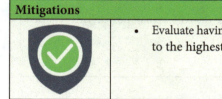	Although you're using avoidance as a way of reducing risk, blocking based on geographic location associated with an IP isn't an effective measure because it can be circumvented. The regulations also take into account whether European customers are targeted by advertising even if their IP is blocked. The CCPA talks about residency status and not your physical location. If you are handling an EU citizen's or Californian resident's data, you need to respect their rights irrespective of where they are physically. Also, more and more nations or states are now implementing similar regulations.
GDPR	Chapter 1, Art. 3 – 1
	Chapter 1, Art. 3 – 2
	Chapter 1, Art. 3 – 3
CCPA and CPRA	*CCPA 1798.140. Definitions (i)*
OECD	N/A

Mitigations	
	• Evaluate having a common baseline of data privacy criteria for all customers to the highest standard, irrespective of heritage

Ace of minimization

You have found a piece of personal data that we can technically do without.

Threat	
	You're collecting data that isn't relevant to the purpose of your activity; this is just exposing you to greater risk for no additional gain. *Chapter 2, Article 5. 1 (c), Data Minimization*, of the GDPR states that collection should be limited to what is necessary for the purpose it was intended.
GDPR	Chapter 2, Art. 5 – 1 (c)
CCPA and CPRA	*CCPA 1798.100. General Duties of Businesses that Collect Personal Information (a)(1)*
OECD	*Part 2, 7. Collection Limitation Principle*

Mitigations	
	• Familiarize yourself with privacy regulations • Don't collect data you don't need; if you don't have permission to use it, you can't anyway • Additional data means additional risk • Make sure what you collect has a specific business purpose

Summary

This was the last of the TRIM categories; the threats covered on the cards were from the Minimization suit of the Elevation of Privacy extension card deck. These threats look at the usage of the data, its relevance, and limiting the amount of data collected to only that absolutely necessary to perform the service or task it was supplied for.

Throughout the book, we have covered threats from the different categories of STRIDE, privacy, and TRIM. The threats on the cards we have covered are not an exhaustive list of all possible threats, but they are some of the more common ones, and understanding them will help you enormously when threat modeling. You should also have a better understanding of the privacy regulations and, as a consequence, your own rights as well. Threat modeling should be performed during the requirements and design phase and may save you costly rework later, but don't forget that a threat model is also a living document that should be updated regularly as the architecture of your software evolves. The objective is to find as many threats as you can and reduce the risk to a reasonable minimum. It's not possible to reduce the risk to zero, and it may also not be cost-effective, but mitigations can reduce the risk, leaving you with a manageable/acceptable level of residual risk.

There are new threat modeling games coming out all the time now as threat modeling becomes more popular. I would suggest mastering one game with or without its extensions first and familiarizing yourself with the threats it aims to expose. Once you have mastered that game, experiment with the others and find the one that works best for you and your team. Some games are tailored more to web apps, some more to cloud and serverless, and some are purely related to privacy. Don't stop there either; you can write your own cards when you see gaps that aren't covered already.

This was also the last chapter of example threats in the book, and I hope you will now have the knowledge you need to threat-model your software designs with confidence. Remember that it's better to start threat modeling early because you're trying to find design flaws and build security into your products from the onset. Happy threat modeling!

Glossary

ABAC - Attribute-Based Access Control.

ACL - Access Control List.

ARP - Address Resolution Protocol.

CAPEC - Common Attack Pattern Enumeration and Classification.

CAPTCHA - A CAPTCHA is a type of Turing test (a test devised by Alan Turing) where you are presented with an image that only a machine should not be capable of comprehending to verify that you are not a robot.

CCSP - Certified Cloud Security Professional.

CI/CD - Continuous Integration / Continuous Delivery.

CISSP - Certified Information Systems Security Professional.

CRLF - Carriage return line feed – in the days of mechanical typewriters, you had to move down one line and then back to the start and this is where this comes from. On Windows, you still see them paired up, however, on Unix-based systems, you just have a newline character or line feed.

CSRF - Cross-Site Request Forgery.

CSSLP - Certified Secure Software Lifecycle Professional.

CWE - Common Weakness Enumeration.

DAC - Discretionary Access Control.

DLP - Data Loss Prevention.

DNS - Domain Name Service.

DNSSEC - Secure Domain Name Service.

EoP - Elevation of Privilege.

GDPR - General Data Protection Regulation (REGULATION (EU) 2016/679).

Geo-fencing - Making a decision based on the geographic location of the person or device.

HIDS - Host Intrusion Detection System.

IAM - Identity and Access Management.

IdP - Identity Provider.

IP - Internet Protocol address – this is a type of network address space. There are two versions in use: IPv4 and IPv6. Generally, when we refer to IPs, we are referring to IPv4 addresses.

ISC2 - International Information System Security Certification Consortium.

JNDI - Java Naming and Directory Interface.

JSON - JavaScript Object Notation.

JWT - JSON Web Token.

MAC - Mandatory Access Control.

MAC - Message Authentication Code.

MFA - Multi-Factor Authentication is when, to log in, you require elements from more than one of the following categories: something you are; something you know; something you have.

mTLS - Mutual Transport Layer Security is the protocol used to secure HTTP but where both the server and the client are required to verify the certificate and identity of one another.

NIDS - Network Intrusion Detection System.

NTP - Network Time Protocol is a network of atomic clocks for synchronizing everyone's computers' clocks.

OAuth - Open Authorization.

OECD - Organization for Economic Co-operation and Development.

OWASP - Open Web Application Security Project is a non-profit organization that endeavors to improve the security of software everywhere through the publication of tools and educational material.

PAM - Privileged Access Management.

Phishing - Tricking people into believing a communication has come from some source they might normally interact with, such as their bank – usually, as an attempt to extort information such as their login credentials or personal data.

PII - Personally Identifiable Information such as names, addresses, telephone numbers, dates of birth, and tax/pension/health identification numbers.

PKI - Public Key Infrastructure.

Rainbow Tables - A dictionary attack using pre-computed hashes or encrypted values.

RBAC - Role-Based Access Control / Rule-Based Access Control.

S.T.R.I.D.E. - A threat modeling methodology that divides threats into six categories represented by the initials. The categories are Spoofing, Tampering, Repudiation, Information Disclosure, Denial of Service, and Elevation of Privilege, respectively.

S.T.R.I.P.E.D. - The S.T.R.I.D.E. threat modeling methodology with the addition of the Privacy category.

S/Mime - Secure/Multipurpose Internet Mail Extensions.

SAML - Security Assertion Markup Language.

SDLC - Software Development Lifecycle.

SIEM - Security Information Event Management.

SMTP - Simple Mail Transfer Protocol.

Social Engineering - The manipulation of humans by preying on their fears.

SSL - Secure Socket Layer is a deprecated/vulnerable protocol for encrypting data in transit using public key asymmetric encryption.

TLS - Transport Layer Security – the evolution of SSL.

TTL - Time to Live.

Vishing - Vishing is the voice version of phishing, tricking people verbally – usually, telephonically.

VPN - Virtual Private Networks are a means by which you can create a tunnel or secure connection that goes across an untrusted network.

WAF - Web Application Firewall.

ZAP - Zed Attack Proxy ZAP is a web app scanner used by penetration testers to find vulnerabilities in web applications.

Further Reading

Books and papers

- Shostack, Adam. "Elevation of privilege: Drawing developers into threat modeling." 2014 USENIX Summit on Gaming, Games, and Gamification in Security Education (3GSE 14). 2014. `https://www.usenix.org/conference/3gse14/summit-program/presentation/shostack`

- Shostack, Adam. "Threat Modeling: Designing for Security", Wiley `https://a.co/d/0jBMmPu`

- Shostack, Adam. "Threats: What Every Engineer Should Learn from Star Wars", Wiley `https://amzn.to/3u0dj81`

- Scott Rogers, Your Turn!: The Guide to Great Tabletop Game Design, Wiley, 2023 `https://a.co/d/izx72Fw`

- Tarandach, Izar and Coles Mathew J. "Threat Modeling: A Practical Guide for Development Teams" O'Reilly `https://a.co/d/cTrhgrS`

- Schoenfield, Brook. "Securing Systems: Applied Security Architecture and Threat Models", CRC Press, 2015 `https://www.amazon.com/Securing-Systems-Applied-Security-Architecture-ebook/dp/B0CVSF9DT8`

- Allen, Christopher and Appelcline, "Shannon. Meeples Together: How and Why Cooperative Board Games Work". Gameplaywright. `https://a.co/d/chIPAdK`

- Johnsson, Deogun and Sawano. Secure by Design, Manning `https://a.co/d/anmMPU7`

- Kohnfelder, Loren. Designing Secure Software, No Starch Press `https://a.co/d/anmMPU7`

- Bell, Bruntonn-Spall, Smith and Bird. Agile Application Security, O'Reilly `https://a.co/d/anmMPU7`

- OECD Council. OECD Guidelines on the Protection of Privacy and Transborder Flows of Personal Data. 2013. `https://bit.ly/3UAEYr6`

- THE EUROPEAN PARLIAMENT and THE COUNCIL OF THE EUROPEAN UNION. REGULATION (EU) 2016/679 OF THE EUROPEAN PARLIAMENT AND OF THE COUNCIL. 2016. `https://eur-lex.europa.eu/eli/reg/2016/679/oj`

Games

- The White Box by Jeremy Holcomb (A kit for creating your own games) `https://atlas-games.com/product_tables/AG2903`

- LINDDUN GO by `https://linddun.org/` (a privacy threat modeling game with a lean approach).

- PLOT4ai – Privacy Library of Threats 4 Artificial Intelligence by Isabel Barberá `https://plot4.ai/`

- Building the Hydrone by Siemens (Teaches the principles of a Secure Software Development Lifecycle) `https://www.thegamecrafter.com/games/building-the-hydrone`

- Data Heist by Mark Barnabas Lee (Data Hygiene and Data Protection awareness game) `https://agilestationery.com/products/data-heist-cyber-hygiene-and-data-protection-game`

- Backdoors and Breaches (a game for incident response tabletop exercises) `https://www.blackhillsinfosec.com/projects/backdoorsandbreaches/`

- Hacker Cybersecurity Logic Game from ThinkFun (A red/blue team game for different skill levels) `https://www.thinkfun.com/products/hacker/`

- @d0x3d (a network security game) from `http://d0x3d.com`

- Careem and TNG Technology Consulting. An online multiplayer version of the threat modeling card games: **Elevation of Privilege** (**EoP**), OWASP Cornucopia and Cumulus. 2019. `https://github.com/tng/elevation-of-privilege`

- TNG Technology Consulting. Cumulus. Threat modeling the Clouds. 2023. `https://github.com/TNG/cumulus`

- Marko Hämäläinen; Laura Noukka; Hiski Ruhanen; Ilona Varis; Antti Vähä-Sipilä of F-Secure Corporation. F-Secure T.R.I.M. Extension. 2018. `https://github.com/WithSecureOpenSource/elevation-of-privacy`

- OWASP Foundation and the Cornucopia Project Team. OWASP Cornucopia is a card game used to help derive application security requirements during the software development life cycle. 2013. `https://owasp.org/www-project-cornucopia`

- Adam Shostack. Adam Shostack's Github Repo. 2009. `https://github.com/adamshostack/eop`

- Mark Vinkovits. Mark Vinkovits Privacy Extension. 2018. `https://logmeincdn.azureedge.net/legal/gdpr-v2/eop-cards-ready-to-print.pdf`

Licenses for third party content

CAPEC:

LICENSE

The MITRE Corporation (MITRE) hereby grants you a non-exclusive, royalty-free license to use Common Attack Pattern Enumeration and Classification (CAPEC™) for research, development, and commercial purposes. Any copy you make for such purposes is authorized provided that you reproduce MITRE's copyright designation and this license in any such copy.

DISCLAIMERS

ALL DOCUMENTS AND THE INFORMATION CONTAINED THEREIN ARE PROVIDED ON AN "AS IS" BASIS AND THE CONTRIBUTOR, THE ORGANIZATION HE/SHE REPRESENTS OR IS SPONSORED BY (IF ANY), THE MITRE CORPORATION, ITS BOARD OF TRUSTEES, OFFICERS, AGENTS, AND EMPLOYEES, DISCLAIM ALL WARRANTIES, EXPRESS OR IMPLIED, INCLUDING BUT NOT LIMITED TO ANY WARRANTY THAT THE USE OF THE INFORMATION THEREIN WILL NOT INFRINGE ANY RIGHTS OR ANY IMPLIED WARRANTIES OF MERCHANTABILITY OR FITNESS FOR A PARTICULAR PURPOSE.

CWE:

Terms of Use

CWE™ is free to use by any organization or individual for any research, development, and/or commercial purposes, per these CWE Terms of Use. Accordingly, The MITRE Corporation hereby grants you a non-exclusive, royalty-free license to use CWE for research, development, and commercial purposes. Any copy you make for such purposes is authorized on the condition that you reproduce MITRE's copyright designation and this license in any such copy. CWE is a trademark of The MITRE Corporation. Please contact cwe@mitre.org if you require further clarification on this issue.

DISCLAIMERS

By accessing information through this site you (as "the user") hereby agrees the site and the information is provided on an "as is" basis only without warranty of any kind, express or implied, including but not limited to implied warranties of merchantability, availability, accuracy, noninfringement, or fitness for a particular purpose. Use of this site and the information is at the user's own risk. The user shall comply with all applicable laws, rules, and regulations, and the data source's restrictions, when using the site.

By contributing information to this site you (as "the contributor") hereby represents and warrants the contributor has obtained all necessary permissions from copyright holders and other third parties to allow the contributor to contribute, and this site to host and display, the information and any such contribution, hosting, and displaying will not violate any law, rule, or regulation. Additionally, the contributor hereby grants all users of such information a perpetual, worldwide, non-exclusive, no-charge, royalty-free, irrevocable license to reproduce, prepare derivative works of, publicly display, publicly perform, sublicense, and distribute such information and all derivative works.

The MITRE Corporation expressly disclaims any liability for any damages arising from the contributor's contribution of such information, the user's use of the site or such information, and The MITRE Corporation's hosting the tool and displaying the information. The foregoing disclaimer specifically includes but is not limited to general, consequential, indirect, incidental, exemplary, or special or punitive damages (including but not limited to loss of income, program interruption, loss of information, or other pecuniary loss) arising out of use of this information, no matter the cause of action, even if The MITRE Corporation has been advised of the possibility of such damages.

Index

packtpub.com

Subscribe to our online digital library for full access to over 7,000 books and videos, as well as industry leading tools to help you plan your personal development and advance your career. For more information, please visit our website.

Why subscribe?

- Spend less time learning and more time coding with practical eBooks and Videos from over 4,000 industry professionals

- Improve your learning with Skill Plans built especially for you

- Get a free eBook or video every month

- Fully searchable for easy access to vital information

- Copy and paste, print, and bookmark content

Did you know that Packt offers eBook versions of every book published, with PDF and ePub files available? You can upgrade to the eBook version at packtpub.com and as a print book customer, you are entitled to a discount on the eBook copy. Get in touch with us at customercare@packtpub.com for more details.

At www.packtpub.com, you can also read a collection of free technical articles, sign up for a range of free newsletters, and receive exclusive discounts and offers on Packt books and eBooks.

Other Books You May Enjoy

If you enjoyed this book, you may be interested in these other books by Packt:

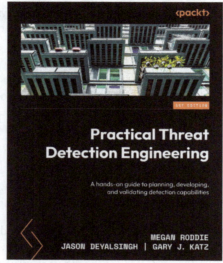

Practical Threat Detection Engineering

Megan Roddie, Jason Deyalsingh, Gary J. Katz

ISBN: 978-1-80107-671-5

- Understand the detection engineering process
- Build a detection engineering test lab
- Learn how to maintain detections as code
- Understand how threat intelligence can be used to drive detection development
- Prove the effectiveness of detection capabilities to business leadership
- Learn how to limit attackers' ability to inflict damage by detecting any malicious activity early

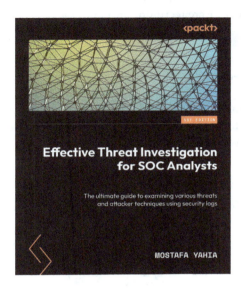

Effective Threat Investigation for SOC Analysts

Mostafa Yahia

ISBN: 978-1-83763-478-1

- Get familiarized with and investigate various threat types and attacker techniques
- Analyze email security solution logs and understand email flow and headers
- Practically investigate various Windows threats and attacks
- Analyze web proxy logs to investigate C&C communication attributes
- Leverage WAF and FW logs and CTI to investigate various cyber attacks

Packt is searching for authors like you

If you're interested in becoming an author for Packt, please visit `authors.packtpub.com` and apply today. We have worked with thousands of developers and tech professionals, just like you, to help them share their insight with the global tech community. You can make a general application, apply for a specific hot topic that we are recruiting an author for, or submit your own idea.

Share Your Thoughts

Now you've finished *Threat Modeling Gameplay with EoP*, we'd love to hear your thoughts! Scan the QR code below to go straight to the Amazon review page for this book and share your feedback or leave a review on the site that you purchased it from.

`https://packt.link/r/1804618977`

Your review is important to us and the tech community and will help us make sure we're delivering excellent quality content.

Download a free PDF copy of this book

Thanks for purchasing this book!

Do you like to read on the go but are unable to carry your print books everywhere?

Is your eBook purchase not compatible with the device of your choice?

Don't worry, now with every Packt book you get a DRM-free PDF version of that book at no cost.

Read anywhere, any place, on any device. Search, copy, and paste code from your favorite technical books directly into your application.

The perks don't stop there, you can get exclusive access to discounts, newsletters, and great free content in your inbox daily

Follow these simple steps to get the benefits:

1. Scan the QR code or visit the link below

https://packt.link/free-ebook/978-1-80461-897-4

2. Submit your proof of purchase
3. That's it! We'll send your free PDF and other benefits to your email directly

www.ingramcontent.com/pod-product-compliance
Lightning Source LLC
Chambersburg PA
CBHW080635060326

40690CB00021B/4944